Other books by Gaylord Johnson:

"Hunting With The Microscope,"
"The Star People,"
"The Sky Movies."

by Irving Adler:
"The Secret Of Light."

Discover
the
STARS

GAYLORD JOHNSON

Revised with additions by

IRVING ADLER

A beginner's guide to the science of Astronomy. Illustrated with 10 astronomical photographs and 41 additional explanatory drawings.

SENTINEL BOOKS PUBLISHERS, INC.

NEW YORK CITY

Published by
SENTINEL BOOKS PUBLISHERS, INC.
112 East 19th Street, New York 3, N. Y.

Revised and Enlarged Edition
Copyright May, 1954
by Irving Adler.

Cover Photograph courtesy of Yerkes Observatory.

Original Edition Copyright
1936 by Gaylord Johnson.

CONTENTS

ILLUSTRATIONS

DISCOVER THE STARS

FOREWORD:

WE WONDER ABOUT OUR FLAT WORLD UNDER A STAR-LINED BOWL

YOU have just picked up, out of curiosity, this book written to help you discover astronomy. You glanced at the title and opened to this page. You want to find out how it is possible for anybody to "discover" astronomy at this late date in the progress of science.

It is true that most of us now think of astronomy as a perfected, "high-brow" science in which mathematical specialists somehow tie up observations made with enormous telescopes to algebraic formulae—and then come out with the statement that there are billions of stars spread out as far as our best telescope can see, two billion light-years away.

This makes us ordinary non-mathematical mortals wonder how we can discover anything about such a transcendental science. We can only take what we read on faith—knowing well, in advance, that we could never hope to understand the explanation.

But fortunately there is another astronomy which we can discover—step by step—with the noble joy of understanding—and as our successive discoveries unfold before us, we can get as great thrills from them as we do from the marching story of an exciting drama.

In fact, the story of man's age-long discovery of the truths of astronomy is a drama in which the human mind, the Ego of the race, is progressively freed from the chains of a multitude of illusions. In it the re-discoveries man has gradually made for himself represent the fascinating story in which the human race has dispelled, one by one, the clouds of illusion and ignorance that in the beginning hid the meaning and extent of Man's home in space—his universe.

5

You and I were born in a century when science seems, at first glance, to have left little for the ordinary man to wonder about. Such facts as the spherical form of the earth, the cause of the seasons and the ocean's tides, the motions of the moon and the planets, the chemical composition of the sun and the stars, are known to be thoroughly understood—and are therefore taken for granted by everybody!

We think of these gradual achievements in knowledge as facts which have always been known. Of course we understand these things!

But how many of us could, if asked, take a pencil and draw a diagram which would make clear to some one else just why summer is hotter than winter? Could you? Could you even give the actual proof that our earth is really a sphere rotating on its axis?

You have heard that we get accurate time daily from the stars (via the U. S. Naval Observatory) to check our clocks, which keep sun-time, based on the daily passage of the sun across the sky. Then how do you account for the fact that your watch (no matter how carefully regulated) is right with the sun only four times a year?

What causes the twice-daily tides in the ocean? "Easy!" you reply, "the moon's attraction raises a wave which travels across the ocean as the earth turns."

Very good—but can you tell me why the moon raises two waves—one on the side of the globe toward it and the other on the opposite side of the earth?

You see that it takes only a few requests for precise explanations to make us see that the knowledge which we take for granted is not knowledge at all—and that we know only those things in which we have become deeply enough interested to "discover" them—step by step—for ourselves.

And there is still another test which determines whether or not we actually understand what we have believed. I mean the ability to explain clearly to some one else the facts upon which our belief is based.

This is the test which I must meet in writing this book for you. And you will in turn be called upon to meet it if you try to explain why, for instance, the planet *Venus* is alternately a "morning" and an "evening" star.

This, by the way, was a discovery which ancient science did not make for thousands of years. Originally, and very naturally, man regarded the "evening star" seen after sunset and the "morning star" seen before sunrise as two separate planets. When they were finally identified as one and the same—the planet *Venus*—a major step in astronomy had been taken.

Can you draw a simple diagram to show how the confusion about *Venus* arose and how it was made clear? If not, your mind is ready to discover what a Greek scientist got a big thrill from finding out several hundred years before Christ!

I am sure that I have now said enough about the subject of this book to make it plain that the hobby of astronomy offers you a chance to get a fascinating and continuous pleasure from following in the footsteps of the scientists who have literally created a new universe for themselves and their posterity by discovering the truths about it.

The first of these scientists was the cave-man who first wondered what happened to the sun when it sank below the horizon. The first scientific theory of astronomy was his speculation that the demon of night had eaten the sun, but would be obliged to belch it up again at sunrise.

And so, if we are to rediscover man's actual knowledge of astronomy from the very beginning, we must go back and stand in imagination upon an earth which our senses tell us is flat—and gaze wonderingly up into a sky which is apparently an inverted bowl sprinkled with sparkling points of fire. As a matter of fact, this requires no imaginative effort for average men and women of today—for they are practically all in the mental state of the cave-man (so far as actual understanding of astronomy is concerned) regardless of the fact that they live in the twentieth century.

If you doubt this, and think my statement too broad, you must take my word for it that I have met educated adults who had never noticed that the stars rise in the east and march all night across the sky to set in the west. This was one of the first astronomical discoveries which some cave-man scientist made, so let us begin by rediscovering what the other pioneers who followed him found out about it.

OUR SKY-BOWL TURNS INTO A
TWIRLING UMBRELLA

WE SHALL never know who it was that first had the idea of illustrating the rotation of the starry sky by means of a turning umbrella. It may well have been an ancient Chinese or Egyptian teacher of science, for umbrellas are almost as old as astronomy. But we do know that an umbrella, with a few stars indicated on its inner surface by chalk dots, is probably the best way of visualizing and understanding the discovery that the heavenly bowl makes one complete rotation every twenty-four hours—and likewise every year.

If you have ever tried learning to know the stars with the aid of an ordinary flat star-map, or "planisphere," you will welcome the simplicity of the umbrella star-map. It is simple because it enables you to imitate realistically the actual movement of the sky—and because you will not be confused by being introduced to an unnecessarily large number of stars in the beginning.

We shall begin, like the Creator, with a dark void (our old black umbrella) and create (with chalk) only the stars we actually need to start with. Then it will be simple indeed to complete our star-map in detail—but we shall do this only after we have grasped the essentials of their daily movement.

LEARN TO KNOW A FEW STARS FIRST

Fourteen stars will do to begin with. Take your umbrella and a piece of white chalk. Indicate with large chalk dots the positions of these stars on the inner surface of the fabric. Get them as nearly as possible into the positions shown in the plan of the umbrella in *Figure 1*. The concentric circles drawn at equal intervals around

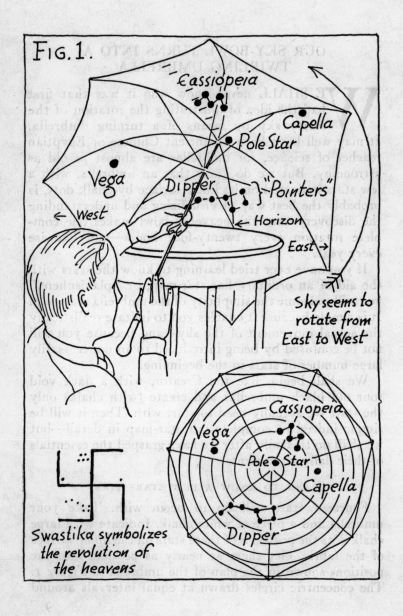

FIG. 1.

Cassiopeia

Capella

Pole Star

Vega

Dipper

}"Pointers"

West

Horizon

East →

Sky seems to
rotate from
East to West

Cassiopeïa

Vega

Pole Star

Capella

Dipper

Swastika symbolizes
the revolution of
the heavens

the point where the umbrella-rod passes through the fabric will aid you in measuring the distances of the stars from the pole around which they all revolve.

Note that there are nine divisions between the umbrella rod and the edge of the umbrella. Each of these represents an interval of 10 degrees on the sky—the entire distance from rod to edge stands for the 90 degree quarter-circle between pole and equator.

A FOUR-SPOKED SKY WHEEL

When you have chalked in the star-groups called the Big Dipper (or *Ursa Major*, the *Great Bear*) and *Cassiopeia*, you will notice that they are roughly at the opposite ends of a line which crosses the pole star. Note also that the two bright stars called *Vega* and *Capella* are likewise at opposite ends of another line, which crosses the pole star at nearly right angles to the first. Accordingly, we may regard these two intersecting lines as forming a wheel of four spokes, which rotates around the pole of the heavens (the umbrella rod). At the outer ends of two opposing spokes are the *Big Dipper* and *Cassiopeia;* at the ends of the other two are the first magnitude stars, *Vega* and *Capella*.

ORIGIN OF THE SWASTIKA

That these four spokes of the heavens have been recognized from the earliest times as forming a wheel of the sky is shown by the symbol of the "swastika"—used the world over as a symbol of the revolution of the heavens around the pole. (See small diagram in *Figure 1*.)

Now that you have completed your umbrella star-wheel of the north, you are ready to see what it will enable you to discover.

Hold the umbrella over a small stand, or a high stool, as shown in *Figure 1*. The angle between the stool top and the umbrella rod should be about 40 degrees, if you live at the latitude of New York or Chicago. At the latitude of New Orleans the angle would be only about 30 degrees. This angle indicates roughly the elevation of the *pole star* above the horizon, which is represented by the far edge of the stool or table top.

DESCRIBE CIRCULAR MOTION THIS WAY

With the umbrella rod held at the appropriate slant to the stool top, place your eye at the level of the "horizon," and turn the rod of the umbrella slowly so that the upper rim of the umbrella moves from right to left. The direction of motion is contrary to the motion of a clock's hands, and is accordingly called "counter-clockwise."

"Counter-clockwise" and "clockwise" are the accurate terms for describing circular motion. The terms "from right to left" or "from left to right," are indeterminate, since any rotating or revolving point travels in both these directions during its circle. "Counterclockwise" makes the direction unmistakable.

Accordingly, when we say that the northern stars travel "counter-clockwise" we mean that, as you face the north, the stars rise in the east, on your right, and pass over toward the west, setting on your left.

Now that the slant of the umbrella rod and the direction of the umbrella's motion are settled, let us see what happens to the stars on the ends of the four "spokes" when we turn the umbrella handle.

STARS WHICH NEVER SET OR RISE

As you rotate it in the direction opposite to the hands of your watch, meanwhile taking care to keep the rod at the correct slant (about 40 degrees) none of the 14 stars you have chalked should disappear below the stool top, with the exception of *Vega*. This bright star will, however, go out of sight for a fraction of its circular course. It will approach the horizon in the northwest, disappear for a few inches, and then rise again in the northeast.

The rest of the 14 stars (the 7 in the *Big Dipper*, the 5 in *Cassiopeia*, and *Capella*) are the chief brilliants in the groups of stars which never set in our North temperate latitudes. Since they travel in circles, round and round the pole, they are called the *"circum-polar"* constellations.

LEARN TO RECOGNIZE GROUPS IN ANY POSITION

As you turn the umbrella, you can become easily familiar with the various positions in which the *Dipper* and *Cassiopeia* appear at different times of night and different seasons of the year. These two constellations are like two children at opposite ends of a gigantic see-saw—when one is up, the other is always down.

It will be well for you to become familiar at once with the way they whirl about the pole at the ends of their opposite spokes, for both the *Dipper* and *Cassiopeia* are not only convenient land-marks from which to locate the other constellations, but can also be used in telling time by the stars, in finding the true north, and in other ways.

PATHS OF EQUATORIAL STARS CURVE SLOWLY

In turning the umbrella as described above, try to keep your eye as near the umbrella rod as possible and as close to the table-edge as you can. With the eye in this position, look up toward the tips of the umbrella's ribs as they pass over your head. Note that the tips travel along a circle so big, that short sections of the circle are almost straight. Then watch the *Pole Star,* near the axis.

13

As you rotate your miniature umbrella-sky, the *Pole Star* makes a small circle around the actual axis—and the stars of the *Dipper* and *Cassiopeia*, as well as *Vega* and *Capella*, all make circles of varying sizes—depending upon their distance from the polar axis, or umbrella rod.

When the ancient observers noticed these details of the sky's westerly movement, they of course believed the motion they saw to be real. Later, however, it became plain that the east to west motion is only an apparent one, caused by the west to east rotation of the earth.

After you have tried this experiment with your umbrella full of stars, you will perhaps be interested to try another one, even more interesting.

HOW TO PHOTOGRAPH TRAILS OF STARS

It is easy to photograph the true rotation of the earth and its effect—the apparent westward motion of the stars. This is the way to do it:

On a bright, clear evening, after dark, set up your camera on a tripod, and tip it up so that you can locate the *Pole Star* at about the center of the finder, or ground glass.

Then focus the camera for infinity (or 100 feet), open the iris diaphragm as far as it will go, set the shutter for a time exposure, and open it. This is all you have to do. You can go away and forget the camera for six hours at least. Before you go to bed, close the lens.

FAST LENSES MAKE BEST TRAIL PICTURES

This experiment succeeds best with "fast" camera lenses (F 4.5 or F 6.3) but I have taken a satisfactory trail picture with an ordinary F 8 "rapid rectilinear" lens.

When the film is developed, you will see many arcs of concentric circles traced in black on the negative. See

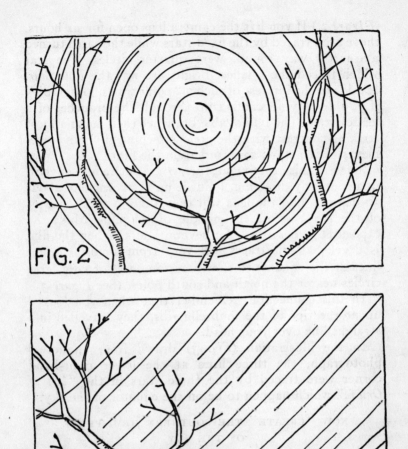

FIG. 2

FIG. 3

(*Figure* 2.) If you left the camera lens open for six hours, these arcs, traced by the fixed stars while the earth turned a quarter way round, were quarter circles. The star which traced the smallest bright arc was the *Pole Star*. It adds to the interest of such a "polar trail photograph" if you point your camera up through the branches of a leafless tree. The immovable branches then make a most striking contrast with the great sweeping circular movements of the stars, as recorded on the film.

To vary this experiment, turn your camera lens upward in a southerly direction. Point it up at an angle of about 50 degrees, and the lens will then take in the "equator of the sky" across the middle of the film. Make the exposure, as described, for several hours. When the film is developed, the traces will vary from almost straight lines at the equator to slowly curving arcs of very large circles nearer the north and south poles. (See *Figure 3*.)

In this connection, remember that when you looked up at the tips of the umbrella ribs, they travelled in a straight line over your head.

In the illustration (*Fig. 3*) drawn from an actual photograph, the three lines at the lower right-hand corner were traced by the three stars in the "*belt of Orion*," which happen to be on the equator of the sky.

NOW "CREATE" THE SOUTHERN HEMISPHERE OF THE SKY

Of course you must remember, in doing these umbrella experiments, that your umbrella represents only the northern half of the celestial sphere—from equator to north pole. In order to complete the picture of the great hollow sphere of stars, you should do one more stunt. To do it, erase the fourteen stars you have chalked in to represent the northern circumpolar constellations.

In their places chalk in some of the stars that are near the south pole: the famous *Southern Cross;* and *Cano-*

pus, a very bright star of the southern heavens. See *Figure 4* for a chart showing where to place the chalk dots.

KEY STARS ROUND THE SOUTH POLE GO "CLOCKWISE"

Now, once again point the umbrella rod upward, holding it upward at about the same angle as before (40 degrees.) The umbrella will then indicate the appearance of a few key stars in the southern heavens, as seen from the latitude of, say, Argentina. As you turn the handle, the stars revolve around the pole in circles, but here is an interesting difference between the experiment you performed with the northern circum-polar constellations and this one with the Southern stars. Instead of turning the umbrella counter-clockwise, you must turn it clockwise. In other words, as you face the south pole, the stars will rise in the east on your left and set in the west on your right. (See *Figure 4*.)

After you have tried this experiment, I suggest that you replace the 14 Northern circum-polar stars with chalk dots as before, as you will need the umbrella later for a very interesting pastime on clear evenings. This will be the making of a more complete star-map of your own, from your own observations, using as your guides a few lines drawn through some of the original 14 stars which are already chalked in your umbrella.

MAKE A PLANETARIUM IN A BOTTLE

While we are on the subject of the celestial sphere, as visualized by your old umbrella, I want to show you how you can get a still more vivid idea of the way the stars rise and set in extreme northern and southern latitudes, and at all points in between. The method involves the construction of a home-made "planetarium" in a bottle.

If you attended the "Century of Progress" in Chicago, you may have seen the wonderful Adler Planetarium in

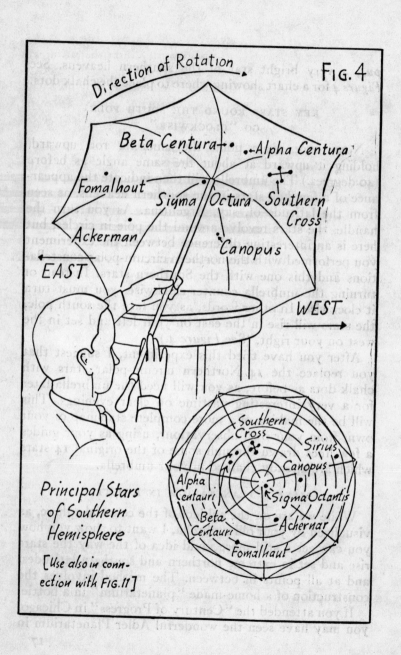

Direction of Rotation

FIG. 4

Beta Centura   Alpha Centura

Fomalhout

Sigma Octura   Southern Cross

Ackerman   Canopus

EAST

WEST

Southern Cross

Sirius

Alpha Centauri   Canopus

Beta Centauri   Sigma Octantis

Achernar

Fomalhaut

Principal Stars
of Southern
Hemisphere

[Use also in conn-
ection with FIG.11]

operation. If so, you saw stars move across the heavens, the moon rise and set, wax and wane, and other interesting phenomena—all in the domed ceiling of the lecture hall. Perhaps you also heard the lecture in which the planetarium illustrated the changing appearance of the sky during a voyage from the north pole to the south pole.

But whether you had the advantage of seeing this wonderful demonstration or not, you can duplicate many of these phenomena with a home-made piece of apparatus.

The essential materials can be secured at any chemist's supply house. They are: a spherical chemist's flask (size unimportant) and a thin glass rod fitting tightly through a perforated rubber stopper which in turn fits into the flask opening. (See *Figure 5*.)

The surface of the flask represents the sphere of stars. The glass rod represents the polar axis round which all the stars apparently revolve.

All that remains to do is to indicate upon the flask's surface, with white paint, the same stars you indicated with white chalk in your umbrella.

MAP YOUR STAR-GROUPS REVERSED

The only difference is that, since your stars are added to the outside of the flask, the arrangement of the stars in the *Dipper*, *Cassiopeia* and any other groups must be reversed—in order that the groups may be seen correctly from the inside of the flask. The diagram (*Figure 5*) shows the position of the stars as placed *in reverse* on the outer surface of the flask.

This *illustration* also shows how any one of the star-groups (the *Dipper*, for example) will be seen the right way round when seen through the opposite side of the glass—which is the nearest you can get to being inside.

19

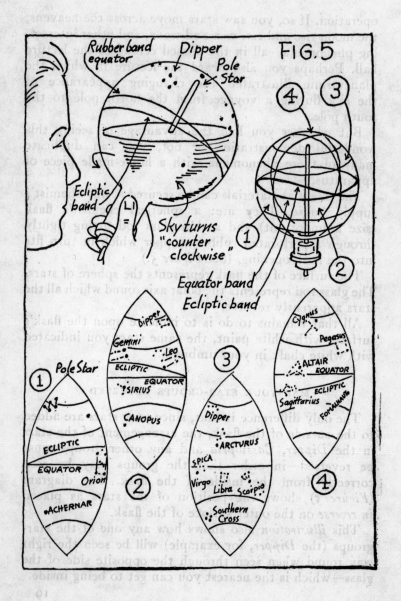

FIG. 5

Rubber band equator

Dipper

Pole Star

Ecliptic band

Sky turns "counter clockwise"

Equator band

Ecliptic band

① Pole Star

W

ECLIPTIC

EQUATOR

Orion

ACHERNAR

② Dipper

Gemini

Leo

ECLIPTIC

EQUATOR

SIRIUS

CANOPUS

③ Dipper

ARCTURUS

SPICA

Virgo

Libra

Scorpio

Southern Cross

④ Cygnus

Pegasus

ALTAIR

EQUATOR

ECLIPTIC

Sagittarius

FOMALHAUT

In locating the various stars marked in the diagram, it is well to mark their places tentatively with a china-marking pencil or other greasy crayon. Then, when all are placed as correctly as possible, the crayon dots can be replaced by dots of white paint.

After the white paint has dried, the planetarium is ready to use in showing the changes that occur in the stars in a voyage from pole to pole. But first we must supply the "ocean."

ALL THE SEA IS INK

The ocean is merely a mixture of water with enough writing ink to give it a dark blue color. Fill the flask half-full of the inky water, after first putting a rubber band around to mark the "equator" or mid-way line between the two poles of the flask.

When the ink is up to this rubber band, you have probably poured in enough, but test it by pushing in the rubber stopper with the glass rod in its center. If, when the flask is held horizontally, the glass rod is just at the surface of the liquid, you are sure that the planetarium is correctly adjusted.

The group of little sketches in *Figure 6* shows the various discoveries which you can make with the ink-filled flask.

HOW PEARY SAW THE POLAR STARS

(*A*) shows how (when the flask is held with the axis vertical) the sky appeared to Admiral Peary when he stood at the north pole. Imagine yourself on a ship at the center of the inky polar ocean. The line where the ink touches the glass all round will then be your "horizon." The *Pole Star* is directly overhead, and as you rotate the flask counter-clockwise (while looking up through its transparent part from below) the stars simply go round and round without rising or setting. If you place a dime under the rubber band to represent the sun, it will also go round and round on the horizon, somewhat as the midnight sun does at one period in the arctic summer. This does not represent the sun's action

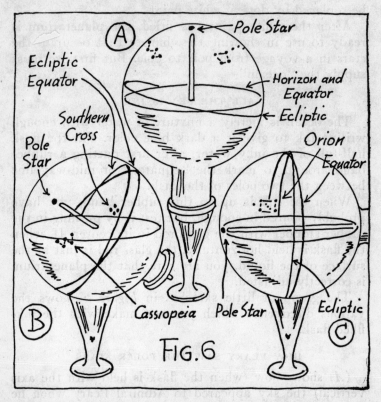

FIG. 6

exactly, as the sun follows the *ecliptic* instead of the *equator*, but it gives the idea roughly.

THE TRAVEL OF THE STARS

(*B*) shows how the stars rise from the sea and set into it in north temperate latitudes. The latitude (as we shall see later in Chapter VI) depends upon the angle between the glass rod and the surface of the sea. This angle becomes smaller as our ship proceeds southward.

WHERE YOU SEE BOTH POLES AT ONCE

(C) shows how, when our ship has reached the earth's equator, the equator of the celestial sphere passes overhead, with the *Pole Star* on the northern horizon and the south pole on the southern horizon.

Be sure to note, when tipping the flask from position (B) to position (C), the point at which the *Southern Cross* rises above the southern horizon.

THE VOYAGE ENDS AT LITTLE AMERICA

If you continue tipping the flask until the neck is above, with the glass rod again vertical, the south pole of the sky will be directly overhead. This indicates the appearance of the stars to the discoverers of the south pole.

If you construct this flask planetarium and carry out this series of experiments with it, you will have a clear idea of the disappearance of the northern stars into the sea, and the appearance of the southern stars from it, during the voyage of a ship or airplane from pole to pole.

SIMPLEST WAYS TO UNDERSTAND STARRY SPHERE

There is no way in which you can so easily re-discover for yourself the mechanism of the daily motion of the starry dome as you can by means of the two devices described in this chapter—the chalk-marked umbrella and stool; and the flask and ink planetarium.

It required hundreds of years, and many thousands of miles of dangerous voyaging, before man arrived at as full a comprehension of his relation to the starry sphere as you are able to gain from a little pleasant experimenting with these simple materials.

And the possibilities of the flask-planetarium are by no means used up; we shall make use of it again (in Chapter IV) to explain the movements and position of the sun and moon during the various seasons.

But it will be more convenient now to re-discover how man learned to tell time by the stars. To accomplish this purpose we shall indulge in a bit of magic—we shall transform our old chalk-marked umbrella into the great star clock of the north. This done, we shall learn how to use it in telling time within a few minutes of watch-accuracy.

II

WE ARE GIVEN A GREAT STAR CLOCK IN THE NORTH

IN ANCIENT, and even in comparatively modern times, the changing positions of certain star-groups in the northern sky were regularly used as indicators to mark the passage of the night. In fact, it is only since clocks and watches have become plentiful and cheap that common people have neglected, and gradually forgotten, the art of telling time by the stars.

In Shakespeare's time, when a modern clock (even if it had existed) could have been only the prized possession of a king or queen, everybody else relied upon the star-clock of the skies to tell the hours through the night—just as they relied upon sun-dials to give the time by day.

TIME TOLD BY THE BIG DIPPER

This is shown by the conversation between the two wagoners in Shakespeare's play "King Henry the Fourth." One wagon driver says to the other:

"Heigh-ho! an' it be not four by the day, I'll be hanged; Charles' Wain is over the new chimney, and yet our horse is not packed."

The meaning is clear at once when we learn that the term "Charles' Wain" is the English equivalent of the *Big Dipper*. The wagoner was thinking of a definite hour of the night when the *Big Dipper* was over the new chimney—four o'clock at the season of year during which he was speaking.

But, even though the art of telling time by the stars is no longer needed, due to our modern profusion of timepieces, it still remains an interesting accomplishment for the amateur astronomer, and may even occasionally become a useful convenience.

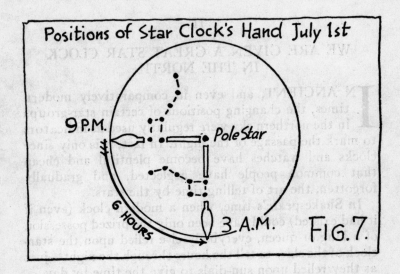

Positions of Star Clock's Hand July 1st

9 P.M.

Pole Star

6 HOURS

3 A.M.

FIG. 7.

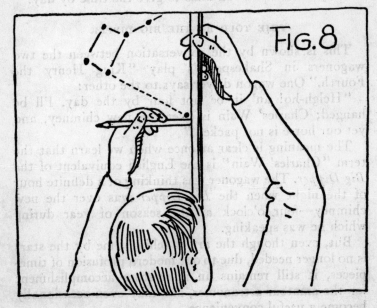

FIG. 8

Suppose, for instance, that you are sleeping out under the stars in a vacation camp. Place the date, for example, on a night near July 15th. You have been asleep on your cot—but awake suddenly and wonder what time it is.

You look at your radium wrist watch, but a glance tells you that you have forgotten to wind it. It does not tick. It has run down. You can only guess that the time is somewhere between midnight and dawn.

You glance up at the brilliantly sparkling vault of the sky. Since your watch has stopped, why not try to rediscover the secret of telling the time of night by the stars? It ought not to be so difficult, since you know that the entire celestial sphere revolves once in about 24 hours.

What can you use for the "hand" of your star clock?

THE STAR-CLOCK HAS A READY-MADE HAND

You remember that on the previous evening you were showing one of the little boys in the camp how to find the *North Star* (or *Pole Star*) by drawing a line through the "pointer stars" in the bowl of the *Big Dipper*. No matter what position the dipper-group of stars assumes at different hours of the night, the "pointers" will always indicate the *Pole Star*. (See *Figures 7, 8,* and *9.*)

Fine! Then why not use the line from the "pointers" to the *Pole Star* as the "hand" of your star clock?

So far, so good. Now if you can only manage to remember what the position of the *Dipper* was last evening about nine o'clock (when you were showing the "pointers" to that youngster) you might be able to estimate the time of night *now*.

TELLING TIME IS SIMPLE

You recollect clearly that the line from pointers to *North Star* was horizontal. The pointers were straight

27

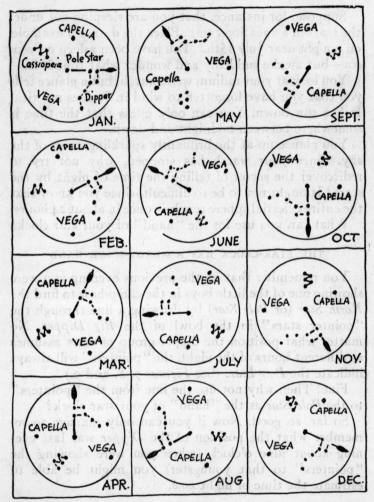

Fig. 9 Nine O'clock Positions of Star Clock's Hand at the middle of each month. Also Two hourly changes of Dipper, Cassiopeia, Vega and Capella.

at the *left* of the *Pole* about 9 p.m. last evening. (See *Figure 7*.)

And a glance at the sky now, at this doubtful hour, shows that the line of the pointers and the *Pole Star* (your clock's hand) are almost exactly vertical. The hand has turned through a quarter circle—or a quarter of 24 hours—or 6 hours. It is accordingly 9 p.m. plus 6 hours, or 3 o'clock in the morning. You have re-discovered the secret of telling time by the stars!

STAR CLOCK HAS TWENTY-FOUR DIVISIONS

Do not get the dial of your great star-clock in the north confused with the dial of an ordinary clock or watch, which is divided into only 12 hourly divisions. The star clock's hand goes only once round the dial in 24 hours, which you must accordingly imagine as divided into 24 hourly parts—6 to each quarter of the complete circle.

A method which will enable you to estimate with considerable accuracy the hours and fractions in this six-hour segment is as follows: (See *Figure 8*.)

HOW TO ESTIMATE ANGLES ACCURATELY

Hold a cane or ruler between your eyes and the sky, letting it hang loosely from your fingers so that it will be exactly vertical, in such a position that its edge seems just to touch the *Pole Star*. Then hold a pencil at right angles to the cane so that the *Pole Star* seems to rest at the tip of the angle formed by the cane and pencil. This makes it easy to estimate the slant of the line from the *Dipper's* pointers to the *Pole Star*—and to estimate how much the observed position of this line (the clock's hand) differs from the position of the hand at some definite hour during that month—at 9 p.m., let us say.

To be properly equipped for telling time by the *Dipper's* pointers and the *Pole Star*, you should be familiar with the 12 different monthly positions of the clock's hand at 9 p.m. The diagram of the star clock in *Figure 9* gives you these positions very simply, and, in addition, the positions at the same times of *Cassiopeia*, *Vega* and *Capella*. By noting the position of *Cassiopeia* (in addition to the *Dipper*) you can learn to use a "hand" extending from the *Pole Star* to *Cassiopeia*, instead of the "hand" extending from the *Pole Star* to the *Dipper's* "pointers."

It is sometimes more convenient to do this, particularly in latitudes where the *Dipper* goes below the horizon when in its position under the *Pole Star*. When either the *Dipper* or *Cassiopeia* is too low for observation, use the group that is high in the sky.

PRACTISE MAKES PERFECT

A good way to become familiar with the different 9 o'clock monthly positions of the circumpolar constellations is to copy off *Figure 9* on a card and carry it in your pocket for a few weeks. Every time you are out of doors on a clear evening, estimate the time from the star clock, and then verify your calculation by consulting your card under the next street-lamp. With a little practise of this kind, you can soon gain an ability and accuracy at time-telling which will surprise your friends.

HOW TO REGULATE YOUR WATCH BY A STAR

One of the things which everybody knows (without understanding it) is that we get our time daily from the stars. But one of the things which almost nobody knows is that any one can use a star to regulate a clock or watch—with no apparatus whatever, except a distant building and a pin stuck into a window sash. (See *Figure 10.*) The method is, however, roughly the same as that used by the astronomer, with his elaborately-accurate zenith telescope, or meridian circle.

FIG. 10

The astronomer knows that a star which crosses the meridian at exactly nine o'clock tonight will come to the meridian tomorrow night at 3 minutes and 56 seconds earlier, or at 8 hours 56 minutes and 4 seconds. We shall find out the reasons and consequences of this fact in the following chapter. It is sufficient now to accept it as a fact.

HOW OBSERVATORY CLOCK IS CHECKED UP

Accordingly, if the astronomer finds that his observatory clock says that the star reaches the meridian at only 3 minutes 55⁹⁄₁₀ seconds earlier than the time of meridian passage on the previous night, he knows that the clock has gained a tenth of a second. He accordingly makes the correction of a tenth of a second in the time signals sent out over the radio.

As already stated, you do not need the elaborate apparatus of an observatory to regulate a watch by the stars in this way. To do it, proceed simply as follows:

HOW TO CHECK YOUR WATCH

Select some star which goes out of sight behind a tall building or a distant house as the star goes down in the western sky.

To observe its disappearance accurately, stick a pin into the cross-bar of your window sash and use the pin head as a "back-sight." In other words, get the pin head and the edge of the building in line with your eye until the star vanishes. As this happens, note the time accurately on the watch you wish to regulate.

Then simply repeat the operation just as carefully on the next evening, and note how much your watch is ahead or behind the 3 minute and 56 second interval.

If your watch says that the star vanishes at 3 minutes and 54 seconds earlier than it did the previous night, your watch has gained two seconds. You must push the regulator a little over toward "slow," and try again the next evening with the same star.

But on the other hand, if your watch records the disappearance of the star behind the building at 3 minutes and 58 seconds earlier than on the previous night, you will know that your watch is *losing* 2 seconds in 24 hours, and must be regulated by shifting the regulator a little toward *fast*.

You can see that if any one star arrives at your line of sight at 3 minutes and 56 seconds (roughly 4 minutes) earlier each night, that all the stars will rise that much earlier night after night. (For directions for making a sun-clock or sun-dial see pg. 115).

CLOCK BECOMES CALENDAR!

During a month of 30 days how much will the sky change through this gradual advance? Four minutes

32

times 30 days is 120 minutes, or 2 hours. And the distance of the sky's advance in 2 hours is one twelfth of the entire revolution of 24 hours. It is apparent then that the sky changes by one-twelfth each month—or by twelve-twelfths (or the entire sphere) in twelve months.

The surprising result is that our star clock has gradually turned into a star-calendar.

In the next chapter we shall re-discover some of the consequences of this interesting transformation.

III

OUR STAR-CLOCK TURNS
INTO A CALENDAR

THIS interesting change was completed at the end of the last chapter when we learned that each month a new one-twelfth of the entire starry sphere comes into view above the eastern horizon—and another twelfth simultaneously sinks below the western horizon.

Accordingly, the stars which come to the meridian (the line joining North and South midway between the horizons) at a certain hour of the evening (say 9 p.m.) will be different every month. They will also, of course, be slightly different every week and every day. However, if we take the interval at every two weeks, we shall be able to construct a star-calendar of the year which will enable us to locate any star-group easily at any time throughout the entire twelve months. The calendar must, however, be made in two parts. (See *Figure 11*.)

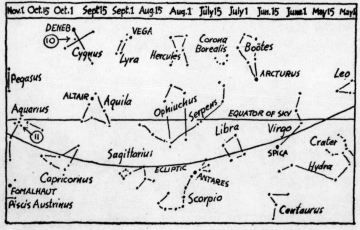

The circular part of Figure 11 maps the circumpolar constellations—the stars which never set in the latitude of the U.S.... (the circular map, turn this book...) when... observing is at the top,... constellation nearest... will then be on the... From the position of the Dipper on... map, and... they correspond with its... position on the clock hand in Figure o in the prev...

EQUATORIAL STAR CALENDAR

The s... map in Fig... possess... part repre- sents the... of the constellations...... the part of the ce... sphere which, to... of... circular circumpol... This map should be imag... as drawn upon a cur... one of... sphere, and... to the circular top.

To use the st...... date, and note the constellations under...... the date... for them at or near the meridian at 8 p.m.....

H...... an hour earlier, at 8 p.m., simply choose... the... earlier... weeks later than your... so... his... of the interval of the... month... the sky... were div... tions of the... revolv... the w... at the... time a clock and a calendar—and the angular advance made in an hour on... clock is about... equivalent to the angular... in two weeks in the calendar. If you find this in any way difficult to... go back to your... ball... with the Dipper and Cassiopeia.

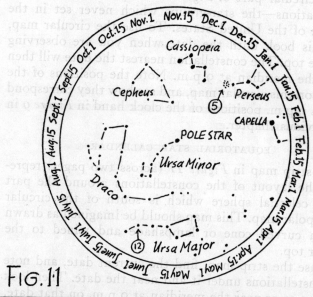

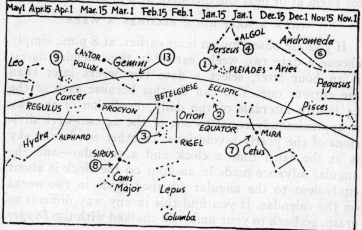

FIG. 11

The circular part of *Figure 11* maps the circumpolar constellations—the star-groups which never set in the latitude of the United States. To use the circular map, turn this book until the date when you are observing is at the top. The constellation nearest the date will then be on the meridian at 9 p.m. Note the positions of the *Dipper* on this circular map, and see how they correspond with its 9 p.m. positions of the clock hand in *Figure 9* in the previous chapter.

EQUATORIAL STAR CALENDAR

The strip map in *Figure 11* (across two pages) represents the layout of the constellations around the part of the celestial sphere which is south of the circular circumpolar map. This map should be imagined as drawn upon a curved cone or lampshade and joined to the circular top.

To use the strip map, find the current date, and note the constellations under it and near the date. Then look for them at or near the meridian at 9 p.m. on that date.

WHEN A HALF-HOUR BECOMES A WEEK!

If you are observing an hour earlier, at 8 p.m., simply choose a date two weeks earlier in the calendar dates; if an hour later, select a date two weeks later than the current one. You can do this because each of the 24 hourly intervals of the daily revolution of the sky is approximately equal to one of the 24 two-week divisions of the yearly revolutions. In other words, the sky is at the same time a clock and a calendar—and the angular advance made in an hour on the clock is about equivalent to the angular advance made in two weeks on the calendar. If you find this in any way difficult to grasp, go back to your umbrella chalked with the *Dipper* and *Cassiopeia*.

36

Place the *Dipper* in its July 15th position — hanging by its handle at the left of the Pole Star. Then rotate the umbrella handle once round, "counter-clockwise," returning the *Dipper* to its same position. A day has elapsed.

Then advance the *Dipper's* position through one twelfth of its circle, and rotate the handle again until the *Dipper* returns to its starting point. A "month" has elapsed between the two "days." You will soon see how the daily and yearly rotations go on simultaneously.

THE DIPPER AS A CONSTELLATION FINDER

Now let us start with the *Dipper* and use it as a finger post to find some other important constellations in the sky.

Hardly any one needs to have the *Big Dipper* pointed out in the sky, but not every one can trace out the complete pattern of stars which forms the *Great Bear*. *Figure 12* shows that the *Dipper* forms only a part of the body and the tail of the bear.

The first direction-line passes through the pointers of the *Dipper* and indicates the *Pole Star* at about the *Dipper's* length above the bowl.

TWO DIPPERS—OR TWO BEARS—TAKE YOUR CHOICE

As *Figure 13* shows, this *North Star* (or *Pole Star*) is the one which marks the end of the handle of another dipper, and the end of the tail of another bear. The two bears (*Ursa Major* and *Ursa Minor*) are indeed very remarkable animals, as far as their tails are concerned. No species at present on earth has a tail even remotely approaching these in length.

While we are considering the two bears (before passing on to find other constellations with their help) let us learn the easy and convenient system which has been devised for labelling each star in a constellation.

37

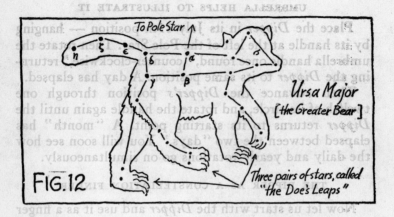

To Pole Star

Ursa Major
[the Greater Bear]

FIG. 12

Three pairs of stars, called
"the Doe's Leaps"

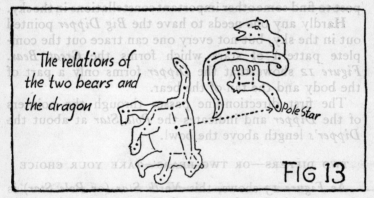

The relations of
the two bears and
the dragon

Pole Star

FIG 13

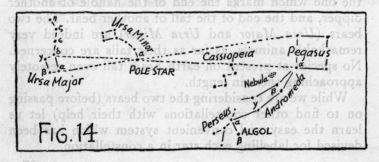

Ursa Minor

Ursa Major

POLE STAR

Cassiopeia

Pegasus

Nebula

Perseus

Andromeda

ALGOL

FIG. 14

The "labels" used for this purpose are the letters of the Greek alphabet. Each star in a group, usually beginning with the brightest in the group, is assigned a letter. In the case of the *Big Dipper,* however, the letters are assigned to the stars in the order of their positions. These letters are very useful in locating the directions of imaginary lines, and in locating any nebulae or star-clusters which may be present in the constellation.

The Greek alphabet, with its sounds, goes as follows:

α Alpha	η Eta	ν Nu	τ Tau
β Beta	θ Theta	ξ Xi	υ Upsilon
γ Gamma	ι Iota	o Omicron	φ Phi
δ Delta	κ Kappa	π Pi	χ Chi
ε Epsilon	λ Lambda	ρ Rho	ψ Psi
ζ Zeta	μ Mu	σ Sigma	ω Omega

In the diagram (*Figure 13*) the seven stars of each dipper use the first seven Greek letters, up to and including Eta.

YOU FIND A STAR BY ITS LETTER

Now for an illustration of the convenience of lettering the various stars. If you look sharply just above the star *Zeta* in the *Great Bear* (the one at the bend of the *Big Dipper's* handle) you will see a faint star just above it. The star *Zeta* was called "the horse" by the Arabs, and the faint star just above it was "the rider." The Arab names were *Mizar* (horse) and *Alcor* (rider). The Arabs also had another name for the rider; they sometimes called it *Saidak* (or "the test") because it is a test of good eyesight to be able to see it with the unassisted eye.

39

But we must get on with our celestial geography. The two bears, Big and Little, are waiting to help us locate more constellations by means of imaginary lines.

The next line (See *Figure 14*) is drawn from the star *Delta* in the *Big Dipper* through the star *Alpha* (or the *Pole Star*) in the *Little Dipper*. When prolonged about the same distance beyond the *Pole Star* this line touches the W-shaped group of five principal stars which we already know as *Cassiopeia*.

Before we go on, pause and notice the winding procession of small stars which curves between the two bears; this is the vague constellation called *Draco*, or "the dragon," useful for judging how bright stars are (Page 115).

LOCATING PEGASUS

We will continue by extending the last line about as far beyond *Cassiopeia* as this W-shaped group is beyond the *Pole Star*. When we do this, we shall locate the line which forms one side of the *Great Square of Pegasus*. The other side of this square is formed by extending the line from the *Big Dipper's* "pointers" to the *Pole Star*. When extended as far as the previous line was, it indicates the other side of the constellation *Pegasus*.

ANDROMEDA AND PERSEUS

Figure 14 shows not only *Pegasus* but also the constellation called *Andromeda*, one star of which is common to *Pegasus*. And when we arrive at the end of *Andromeda*, we find ourselves touching the curving row of stars which forms the most characteristic feature of the constellation *Perseus*.

Figure 14 will enable you to locate these groups easily by their relation to each other and to the *Dippers*, big and little.

While we are in the neighborhood of *Andromeda* and *Perseus*, we must pause a moment to locate two features of considerable interest—both of which can be seen with the naked eye.

FINDING THE GREAT NEBULA

The first is the famous *Great Nebula* in *Andromeda,* of which you have undoubtedly seen photographs taken with great telescopes. It is located just above the star *Beta* in *Andromeda* (in the direction of *Cassiopeia*). On clear, moonless nights it is visible to the unaided eye as a tiny wisp of fuzzy light. An opera or field glass brings it out more plainly.

ALGOL, OR "THE DEMON"

The other feature is the celebrated variable star named *Algol.*

The constellation *Perseus* contains a very remarkable star (designated by the Greek letter *Beta* in Figure 14) which varies greatly in brilliancy. Its ancient name is *Algol,* or "the Demon." At its brightest, *Algol* is of the second magnitude—(about the brightness of the stars in the *Big Dipper*). It remains bright for 2½ days. Then it fades for 4½ hours—and becomes an inconspicuous fourth magnitude star—which it remains for a few minutes only. Its increase in brightness also takes 4½ hours—after which the cycle begins again. The interval between times of minimum brightness is 2 days, 20 hours and 49 minutes—so you can calculate its periods ahead for yourself after observing one of them. All the observation of *Algol* for over 200 years goes to prove that a dark companion revolves around *Algol,* almost in the plane of our line of sight, producing as it passes in front of the bright star a partial eclipse of its light.

But now let us return to the *Great Bear,* after this detour to view *Algol,* and see what other star groups this useful animal will help us to locate.

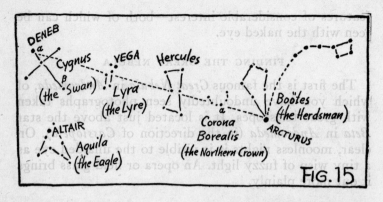

DENEB
α
Cygnus
(the Swan)
β
VEGA
Lyra
(the Lyre)
Hercules
α
Corona
Borealis
(the Northern Crown)
Boötes
(the Herdsman)
ARCTURUS
ALTAIR
Aquila
(the Eagle)

FIG. 15

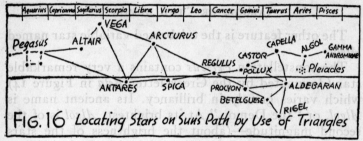

Aquarius	Capricornus	Sagittarius	Scorpio	Libra	Virgo	Leo	Cancer	Gemini	Taurus	Aries	Pisces

Pegasus
ALTAIR
VEGA
ARCTURUS
CAPELLA
ALGOL
GAMMA
ANDROMEDAE
REGULUS
CASTOR
POLLUX
Pleiades
ANTARES
SPICA
PROCYON
BETELGEUSE
ALDEBARAN
RIGEL

FIG. 16 Locating Stars on Sun's Path by Use of Triangles

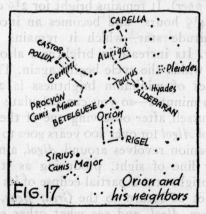

CAPELLA
CASTOR
POLLUX
Gemini
Auriga
Pleiades
Taurus
Hyades
ALDEBARAN
PROCYON
Canis Minor
BETELGEUSE
Orion
RIGEL
SIRIUS
Canis Major

FIG. 17 Orion and his neighbors

Even his tail forms a pointer for a constellation, as you will see by *Figure 15*. If you extend the curving line which joins the three stars in the *Dipper's* handle it will bring you (at about a dipper's length) to the beautiful yellow-white star *Arcturus*.

You will note at once that *Arcturus* is at the foot of a kite-shaped figure. This is the main part of the constellation called "Boötes," or *The Herdsman*. It looks far more like a kite than a shepherd. This figure is also often referred to as the *Bear Driver*, owing to the fact that it follows the *Great Bear* in its ceaseless march around the *Pole Star*.

CORONA BOREALIS

Alongside of the *Herdsman* is a small broken circle or garland of five stars. One of these (marked with the Greek letter *Alpha* in *Figure 15*) is of the second magnitude—the rest are all third or fourth. The whole garland is called *Corona Borealis*, or the *Northern Crown*, a beautiful little constellation, easily found.

Going back to the kite-shaped *Herdsman*, we can use him to help us locate the rest of the circumpolar constellations. All that is necessary is to draw a line from *Arcturus* through the brightest star in the *Northern Crown* and extend it as shown in *Figure 15*. This extended line will be our guide to the four remaining constellations of importance in the northern circular map of *Figure 11*.

The first of these groups crossed by the line from *Arcturus* is *Hercules*, a large but rather vague constellation, with no naked eye features of much interest.

VEGA, BRILLIANT OF THE LYRE

However, after our line has crossed *Hercules*, it brings us to the brilliant blue-white star *Vega*, one of the

43

beauties of the northern sky. *Vega* is the principal star of a compact little group of much fainter stars. This constellation is called *Lyra*, or *The Lyre*, although there is no apparent resemblance to a musical instrument of any kind.

SWAN AND EAGLE

Continuing beyond *Vega* the line which began in the side of the *Herdsman's* kite, it crosses the plainly marked *Northern Cross*, or *Cygnus*, *The Swan*. The bird is imagined as if in flight, with outspread wings along the short arm of the cross. Its tail is marked by *Deneb* (the *Alpha*, or brightest star of the constellation) and its long neck terminates at *Beta*, which locates the foot of the cross.

The relationship of *Aquila*, or *The Eagle*, to the *Swan* is shown by *Figure 15*. The *Eagle* is easy to locate, because it contains the bright, first magnitude star *Altair*.

Having now located the principal star groups in the northern circumpolar group, we must learn the sky geography of the band of sky which surrounds our earth above and below the equator and is consequently called the "equatorial belt." This includes the ancient and famous group of 12 "zodiacal" constellations.

THE ZODIAC—A SKY MENAGERIE

The word comes from a Greek word "zodiakos," meaning an animal, and its application is obvious when you run over the names of the signs, or constellations of the zodiac, which are as follows, in their order from the Spring equinox onward. The symbols are those used in the almanacs.

The Ram ♈; *the Bull* ♉; *the Twins* ♊; *the Crab* ♋; *the Lion* ♌; *the Virgin* ♍; *the Balance* ♎; *the Scorpion* ♏; *the Archer* ♐; *the Goat* ♑; *the Water-Carrier* ♒; *the*

44

Fishes ♓. *The sign* ♈ represents the horns of the *Ram;* ♉ the head and horns of the *Bull,* and so on.

The strip map (across two pages in *Figure 11*) will show you when to look for each of the zodiacal constellations on the meridian—and these star groups are easily recognized. We need only to run over them briefly, calling attention to the chief features of each. In this connection *Figure 16* will be a help in locating the principal stars in each group along the ecliptic.

THE RAM LEADS THE PROCESSION—AND THE BULL FOLLOWS

Aries, or the *Ram,* is the first sign of the zodiac. He is a faint constellation lying between *Andromeda* and the *Pleiades*—and is on the meridian during December.

To the left, or eastward of the *Ram,* you see *Taurus,* the *Bull,* much more conspicuous and splendid to the eye. The head forms a triangle of sparklers in which burns *Aldebaran,* a magnificient reddish star which marks the *Bull's* right eye. The many small scintillating stars on the *Bull's* face and forehead remind one of the *Pleiades,* described by Tennyson as "a swarm of fireflies tangled in a silver braid," and so does this little group's name, which is "the *Hyades.*"

The *Pleiades* themselves are also in the *Bull,* sparkling upon his shoulder. See if you can count more than six stars in the *Pleiades* with the naked eye. The telescope shows several hundred.

TWINS, CRAB AND LION

Next comes the sign *Gemini,* or the *Twins.* It is easily recognized by *Castor* (*Alpha*) and *Pollux* (*Beta*), the two fine first magnitude stars which mark the twins' heads.

The next sign, *Cancer,* the *Crab,* is very inconspicuous to the naked eye, but very interesting for the opera or field glass, as we shall see in Chapter VIII.

The next constellation toward the east is *Leo*, the *Lion*, on the meridian in late March and early April. *Leo's* principal feature is the brilliant first magnitude star *Regulus*, which is very close to the line of the ecliptic, marking out the path of the planets, moon and sun.

VIRGIN, SCALES AND SCORPION

Virgo, or the *Virgin*, also exhibits as her outstanding glory a single bright star, *Spica*, or "the ear of wheat." It is interesting to note in *Figure 16* that *Spica*, with *Regulus* and *Arcturus*, form a large triangle which helps in locating them all at any time.

Libra, the *Balance*, or the *Scales*, is unimportant and inconspicuous, but *Scorpio*, the *Scorpion*, which is next in order, is one of the most beautiful of all the constellations. Also, it is almost the only star group which really looks a little like the object for which it is named.

The chief glory of *Scorpio* is likewise a single first magnitude star, reddish in color, named *Antares*. This name, which means "rival to *Mars*," is explained when we remember that *Ares* is the Greek name for *Mars*, and that this red planet passes near *Antares* in its travels along the ecliptic around the sun. At times of near approach to *Antares*, *Ares* (or *Mars*) might be mistaken for its rival—and vice versa.

ARCHER, SEA-GOAT, WATERMAN AND FISHES

Eastward from the *Scorpion* we come upon *Sagittarius*, or the *Archer*, who always keeps his arrow aimed at *Scorpio's* heart. The archer shows a distinct bow and arrow outlined in stars—also a figure called "the *Milk Dipper*," but no first magnitude stars.

As indicated in *Figure 11*, *Sagittarius* and *Scorpio* are on the meridian in July and August. They are then low in the sky (in north temperate latitudes) and are seen only during the summer months.

46

Capricorn, or the *Sea Goat,* is a faint group with no outstanding brilliants. *Aquarius,* or the *Water-Carrier,* is also not rich in bright stars, but has a multitude of tiny sparklers, which are arranged in streams that suggest flowing water.

Finally, *Pisces,* or the *Fishes,* the concluding sign of the Zodiac, is found just south of the great square of *Pegasus.* It is a diffuse constellation and the long "ribbon" of small stars which runs southward to the *Southern Fish* ends in a bright star called *Fomalhaut.* It is very close to the horizon in north temperate latitudes.

MINOR GROUPS FOUND FROM STAR MAPS

Now that we have made the circuit of the constellations which form "the houses of the sun" upon his annual pilgrimage, you will be able to locate from them any of the smaller and fainter star groups which may attract you. Little by little, and one by one, you will spy out the smaller groups, such as the *Dolphin,* the *Hair of Berenice, Leo Minor* and many others. Any good star-atlas will help you if you decide to learn the constellations thoroughly.

ORION, THE WONDER OF WINTER

We have, however, in following straight through the Zodiac, omitted a constellation which is probably the grandest of them all.

I mean *Orion,* the splendor of the Winter sky. Its relations to neighboring star groups are shown in *Figure 17. Orion* is just off the road of the Zodiac, but it is difficult to see how it escaped becoming one of the signs, as it is so outstandingly magnificent.

In Chapter VIII we shall return to *Orion* and examine some of its beauties at closer range through a field glass.

IV

WE SEE THE PAGEANT OF
THE VARIED SEASONS

IF WE are to re-discover the way in which primitive man gradually explained the many changes in climate and vegetation which we call the seasons, we must place ourselves (in imagination, at least) in the surroundings in which he observed.

He did not, of course, live in cities where the glare of street lights and shop windows, and the smoky atmosphere, prevented him from seeing and noticing the stars and the horizon. Also, clocks and watches did not prevent early man from needing to use the stars to tell his time. Almanacs and calendars in every home did not save him the need of watching the sun's position on the horizon at rising and setting—and the length of shadows cast by objects—in order to know how far advanced the winter was, or when to plant his crops in the spring.

PIONEER INSTRUMENTS

Accordingly, we can infer that two of the first "astronomical instruments" ever used by man were—first, the *horizon* of the place where he lived, and second, a *vertical stick*, or spear, fixed upright in the ground.

If it seems to you an exaggeration to call these simple things "astronomical instruments," consider some of the discoveries which they enabled man to make, and see if you do not agree with me.

In all probability, man's very first astronomical instrument was the far-off horizon of a flat plain, dotted with familiar trees, rocks, and so on.

Primitive man, if observant at all, could hardly watch the sun rise for more than a few days in succession without noticing that its place of first appearance in the morning (as well as its last visible position at night) shifted gradually either northward or southward.

SHIFT OF SUNRISE AND SUNSET POINTS

The places of sunset and sunrise move Northward gradually from the date we now call the "winter solstice" on December 22nd, to the time of the "summer solstice" on June 21st. At this time, after the farthest northward point of sun-rise apparently stood still for a few days, man noticed that the sun's places upon the horizon gradually began to shift southward day after day, and week after week, until they again reached the points from which they started, in the middle of the previous winter.

The diagram (*Figure 18*) shows how early man used his first instrument for studying astronomy, and how you can repeat the discoveries he made with it.

You can see how an observant, or "scientific" savage would watch the sun for only a few years before he would be able to settle the length of the year with considerable accuracy.

INVENTION OF SUNDIAL

And then, undoubtedly, some primitive genius made an important improvement upon the horizon as an astronomical instrument. He based his invention upon the discovery that a spear, stuck upright in the ground, casts a shadow which varies in length during the day, and also throughout the year.

It would be natural for a savage to notice that the shadow shortened as the sun rose higher in the sky every day, and that the shortest shadow always pointed in the same direction in which he saw the motionless *Pole Star* by night. In other words, that the shadow at midday pointed north.

49

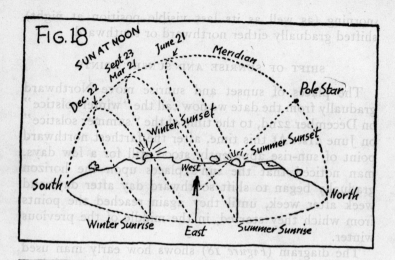

FIG. 18

SUN AT NOON

Dec. 22 · Sept 23 Mar 21 · June 21 · Meridian

Pole Star

Winter Sunset

Summer Sunset

West

South

North

Winter Sunrise · East · Summer Sunrise

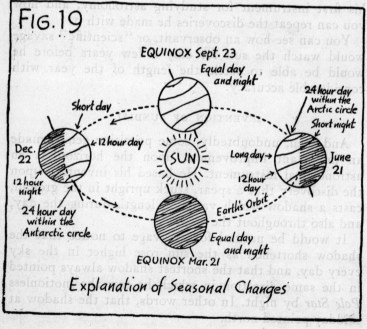

FIG. 19

EQUINOX Sept. 23

Equal day and night

Short day

24 hour day within the Arctic circle

Short night

Dec. 22

SUN

Long day

June 21

12 hour day

12 hour night

Earth's Orbit

12 hour day

24 hour day within the Antarctic circle

Equal day and night

EQUINOX Mar. 21

Explanation of Seasonal Changes

And as early man watched this shadow of his spear (or perhaps a tall, straight pine tree) from month to month, he could not fail to notice that the shadow increased its length gradually, and then shortened it again. A six-foot spear would show a shadow (in latitude 40 degrees) of 12 feet at mid-winter and less than 2 feet at mid-summer. In some such way the first sundial originated, and as civilization increased, the spear took permanent form as the "obelisk" of an Egyptian temple to the sun god.

In this way, when man had noticed that the time of the longest shadow coincided with the time of the sun's farthest southward advance on the horizon—and the shortest shadow with the farthest Northward advance— he discarded the use of the horizon in favor of the more convenient sun dial.

And so, through hundreds and thousands of years, man first built up his knowledge by *observation*, and then used his *reason* to explain why the facts he observed came to be.

Our present day understanding of the differences in the length of shadows throughout the year, as well as the varying lengths of the day, can be shown more clearly in a diagram than in a great many words. A little study of *Figure 19* will make the essentials clear.

SLANT OF EARTH'S AXIS CAUSES SEASONS

The single astronomical fact upon which all the seasonal changes depend is that the axis of the earth is not perpendicular to the plane of its orbit, but is inclined at an angle of 23½ degrees from a perpendicular to the plane. For practical purposes, the axis always indicates the same point among the stars—a point about two degrees from the *Pole Star*.

51

Because of the slant of the earth's axis, the northern and southern hemispheres are alternately inclined toward the sun and away from it.

LONGEST AND SHORTEST DAYS

On June 21st, the inclination of the northern hemisphere toward the sun is at its maximum, and the Arctic circle is in the sunlight for the entire day, and days in the northern hemisphere are at their longest.

On December 22nd, the tilt of the northern hemisphere away from the sun is greatest, and the Arctic circle is out of the sunlight during the entire day, and causes the days in the northern hemisphere to be at their shortest. In the southern hemisphere conditions are of course completely reversed; northern summer is southern winter, and vice versa.

WHY SUMMER IS HOTTER THAN WINTER

Note, however, the interesting fact that at the equator the day and nights are equal in length throughout the year.

The true reasons why summer is a hotter season than winter can be shown easily by the simple experiment shown in *Figure 20*.

Many people think that the summer is a hotter season because the earth is nearer to the sun during June, July and August, yet the fact is that our earth's orbit actually carries it nearest to the sun at mid-winter!

The true reason for summer's warmth and winter's chill lies in the slant of the earth's axis to the plane in which our planet revolves round the sun. The simple experiment pictured in *Figure 20* makes this plain.

SUN'S HEAT STRIKES AT HIGHER ANGLE IN SUMMER

A globe mounted upon an axis inclined twenty-three and one-half degrees to the vertical is turned so that it leans toward a flash light representing the sun. This indicates the earth's position at midsummer. When the

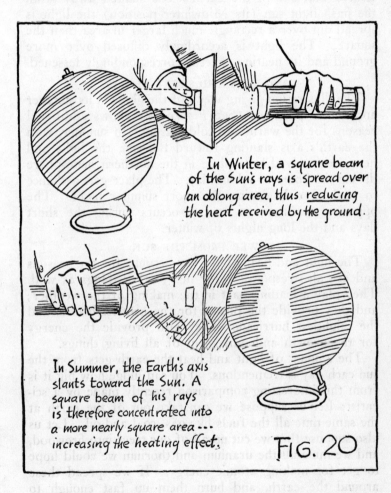

In Winter, a square beam of the Sun's rays is spread over an oblong area, thus _reducing_ the heat received by the ground.

In Summer, the Earth's axis slants toward the Sun. A square beam of his rays is therefore concentrated into a more nearly square area... increasing the heating effect.

FIG. 20

square beam of light passing through a hole in a card falls upon the United States, it is concentrated upon a spot almost square. Its heating effect is therefore intense.

When the identical square beam is turned upon the United States with the earth's axis slanted away from the flash light sun (its Midwinter position) the light is spread out over a rectangle much larger in area than the square. The light is accordingly diffused over more ground and its heating effect is correspondingly lessened.

LONG DAYS AND SHORT NIGHTS

The illustration giving views of the globe in its summer and winter positions (See *Figure 19*) shows additional reasons for the warm and cold seasons. In summer, with the earth's axis slanting toward the sun, the light and heat are received by a place in the northern hemisphere during the hours of a long day. The place gets a chance to cool off only during the short summer night. The opposite condition, of course, occurs during the short days and the long nights of winter.

POWER FROM THE SUN

The light and heat of the sun account for the seasons and are also responsible for many other things as well. They play a leading part in the making of the weather; and they provide the power for the gentlest breeze and the mightiest hurricane. They also provide the energy for the growth and movement of all living things.

The amount of light and heat the earth gets from the sun each day is tremendous. You can judge how big it is from this interesting comparison recently made by a scientist: Let us suppose we can take out of the earth at the same time all the fuels like gas, coal and oil. Let us also suppose that we cut up all of our trees into firewood, and we mine all the uranium and thorium we could hope to get for making atomic energy. If we spread these around the earth, and burn them up fast enough to provide as much energy as we get from the sun, they would last only three days! The energy we get from the sun is so much compared to all other sources, that

54

scientists and engineers are now trying to invent ways to capture it, for use in running our factories and heating homes.

OUR GAIN IS THE SUN'S LOSS

While the energy we get from the sun is big compared to all our other sources, it is small compared to the total amount of energy the sun pours out each day. The sun sends its rays out in all directions, and we get only that tiny part that heads our way. The power sent out by the sun is about five hundred thousand billion billion horsepower. To provide this power the sun uses up part of its mass, and so loses mass at the rate of four million tons a second.

SUNLIGHT AND THE H-BOMB

The process that creates all this energy deep in the interior of the sun is a nuclear reaction that combines hydrogen atoms to make helium atoms. Physicists understand this process so well that they are using it now in the manufacture of H-bombs. The same process is the source of the energy for all those distant suns we see at night, the stars.

V

WE SPEND TWO THOUSAND YEARS LEARNING TO UNDERSTAND THE MOVES OF THE PLANETS

WHEN some ancient shepherd, watching his flocks under the star-glinting dome, first noticed that a few of the stars do not remain fixed in their places, he made a discovery which took man thousands of years to explain with completeness.

Gradually, as the result of generations of persistent observing, the ego of the human race enlarged his knowledge of these steadily moving stars, which were early called "planets," or "wanderers."

PLANETS ALL MOVE EASTWARD

Man soon found that the planets go through certain definite evolutions and then repeat them. The simplest of these movements is their steady eastward progress among the stars of the zodiacal constellations. In course of time the scientific spirits in succeeding generations devised a theory to explain this motion. They decided that the wandering stars, or planets (in which they included the sun and moon) travelled around the earth along the road of the zodiac, eventually returning to their starting points.

This explanation would have proved quite satisfactory if it had not been for the other and much more puzzling motion which the wandering stars were observed to make.

WHY DID PLANETS SOMETIMES GO BACKWARD?

The phenomenon which demanded further explanation was this: The primitive astronomers of early days had found that a planet, after proceeding steadily along the starry path of the zodiac for many months, gradually slowed down, and eventually stopped its travel altogether!

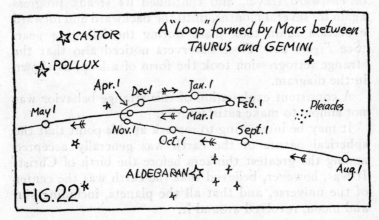

FIG. 22

A "Loop" formed by Mars between TAURUS and GEMINI

☆ CASTOR
☆ POLLUX

Apr. 1 Dec. 1 Jan. 1
May 1 Nov. 1 Mar. 1 Feb. 1
Oct. 1 Sept. 1
ALDEBARAN ☆ Aug. 1

Pleiades

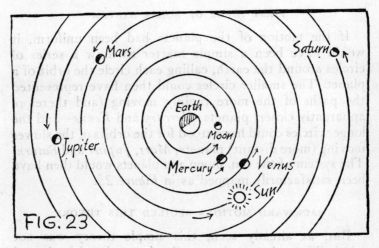

FIG. 23

Mars Saturn
Earth
Jupiter Moon
Mercury Venus
Sun

And then, after remaining apparently stationary for a few days, the planet was observed to travel backward along its road. This motion continued for several weeks, until it also slowed down, and the planet again became

57

stationary. Finally, after another pause of a few days, the wandering star again gradually resumed its forward, or eastward travel, and continued its steady progress again for several months, until the backward and forward evolution again took place during the following year. (See *Figure 22*.) Close observers noticed also that the strange retrogression took the form of a loop, as shown in the diagram.

A consistent explanation of this strange behavior was not simple to make satisfactorily.

It may be interesting to remark at this point that the spherical nature of the earth was generally accepted among the greatest thinkers before the birth of Christ. It was, however, believed that the earth was the center of the universe, and that all the planets, including sun and moon, revolved around it.

FIRST IDEAS OF SOLAR SYSTEM

If the motion of the planets had been uniform, it would have been a simple matter to draw a series of circles around the earth, calling each circle the orbit of a planet. The smaller circles could then have represented the paths of the more rapidly moving (and therefore apparently closer) planets, *Mercury* and *Venus*—and the longer circles could have stood for the orbits of the slower moving (more distant) planets, *Mars*, *Jupiter* and *Saturn*. The system of the sun, moon and planets would then have been satisfactorily mapped as in *Figure 23*.

BACKWARD MOTIONS SPOILED THIS THEORY

But, as already seen, this simple theory was not admissible, owing to the peculiar forward and backward evolution of each planet once during each year. An example will make clear the strange problem which nagged the ancient astronomers for a solution.

The planet *Jupiter* was observed to take twelve years to accomplish its apparent journey completely around the earth. Accordingly, during that circuit, *Jupiter* went through twelve loop-movements, as shown in our diagram in *Figure 24*. As shown there, the sun was supposed to be nearer to the earth than the outer planets!

This problem was without a satisfactory explanation until an astronomer called Ptolemy wrote a book (about 140 A.D.) in which he put forth the following system of the universe. Ptolemy stated that the puzzling "loops" observed to take place in the paths of the planets were really circles seen edgewise.

PTOLEMY'S IDEA OF THE COSMOS

Accordingly, he drew his map of the universe somewhat as shown in *Figure 25*. He supposed that a planet (*Jupiter*, for instance) travelled in small circles called "epicycles," while at the same time moving in a larger circle around the earth, called the "deferent."

Neither Ptolemy nor any one else was able to explain why the planets moved through their epicycles, but when it was granted that they did, the theory explained beautifully all the phenomena which were observed to take place. In fact, Ptolemy's view of the universe became so well established in the world's centers of learning that it was not seriously questioned until the coming of another astronomer named Copernicus, who lived just about the time of the discovery of America by Columbus. The illusion that the sun and planets all revolved around the earth held sway for nearly 1400 years!

COPERNICUS THE REVOLUTIONARY THINKER

Copernicus rejected flatly the idea that the earth was the center of the universe—and, in a revolutionary book,

59

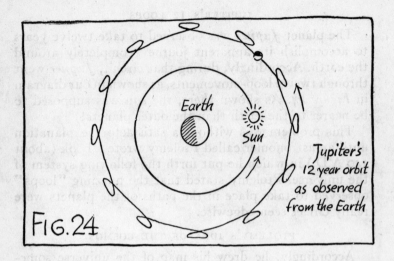

Earth

Sun

Jupiter's
12 year orbit
as observed
from the Earth

FIG. 24

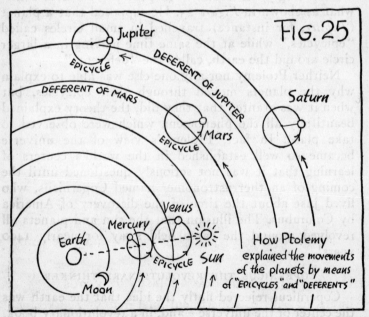

FIG. 25

Jupiter

EPICYCLE

DEFERENT OF JUPITER

DEFERENT OF MARS

Saturn

EPICYCLE

Mars

Venus

Mercury

EPICYCLE

Earth

Sun

Moon

How Ptolemy
explained the movements
of the planets by means
of "EPICYCLES" and "DEFERENTS"

he explained the movements of the planets, moon and sun substantially as we do today. See *Figure 26*, which is a reproduction of an illustration from Copernicus' book—*De Revolutionibus Orbium Celestium* (or the Revolution of the Celestial Spheres).

SLOW TO GAIN ACCEPTANCE

The idea that the orbits of the planets had the sun as center instead of the earth, did not however, come into general acceptance for about a hundred years. In fact,

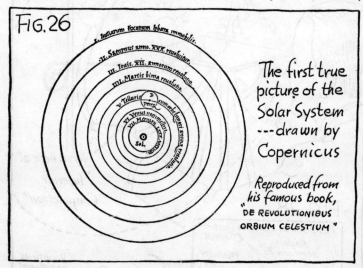

FIG. 26

The first true picture of the Solar System ---drawn by Copernicus

Reproduced from his famous book, "DE REVOLUTIONIBUS ORBIUM CELESTIUM"

amazing to relate, the Ptolemaic system was for a considerable time taught in both the newly established Harvard and Yale colleges in connection with the Copernican! Sort of a "pay your money and take your choice" method of teaching astronomy.

HOW TO DEMONSTRATE PHASES OF VENUS

The simple little experiment pictured in *Figure 27* (together with the diagram in Figure 28) will make plain

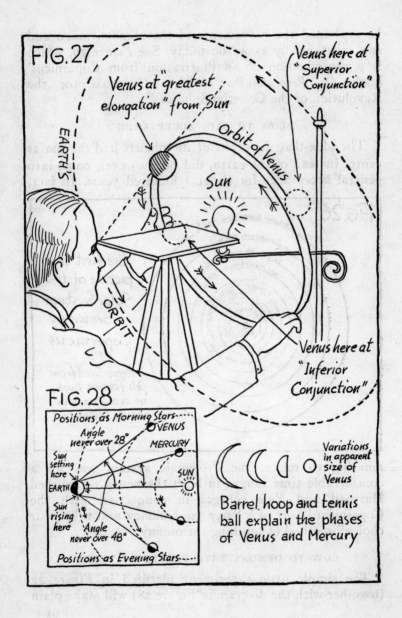

FIG. 27

Venus here at "Superior Conjunction"

Venus at "greatest elongation" from Sun

EARTH'S

Orbit of Venus

Sun

ORBIT

Venus here at "Inferior Conjunction"

FIG. 28

Positions as Morning Stars
VENUS
Angle never over 28°
MERCURY
Sun setting here
EARTH
SUN
Sun rising here
Angle never over 48°
Positions as Evening Stars

Variations in apparent size of Venus

Barrel hoop and tennis ball explain the phases of Venus and Mercury why *Venus* and *Mercury* are sometimes "evening stars" and sometimes "morning stars."

A slit has been cut in an old tennis ball and the edge of a barrel loop forced into it. This supports the ball on the edge of the hoop. The square of cardboard on the camera tripod represents the western horizon. The lighted bulb of the bridge lamp represents the sun, setting on the horizon. When the hoop is turned in the direction of the arrow, the lamp illuminates the ball exactly as the sun lights up *Venus*. The "phases" of Venus can be duplicated in this way—from "crescent" to full and back again. Also the rotation of the hoop, carrying the ball out of line with the lamp and then back into line with it, illustrates how the planet *Venus* appears near the setting sun—gradually travels eastward along the zodiac until it is over 40 degrees from the setting sun, and then "retrogrades" westward until it is again in line with the sun and disappears into the morning sky. There *Venus* appears in the east before sunrise and goes through similar evolutions.

Note in *Figure 28* that *Mercury* follows the same routine as *Venus*.

THE YEARLY RACE OF MARS AND THE EARTH

The formation of the loop in the paths of *Mars*, *Jupiter* and *Saturn* may be better understood by a moment's study of *Figure 29*. Since the earth travels through its orbit in one year, while *Mars* requires two, the earth gains upon and passes *Mars* once each year. At these times, the motion of *Mars* apparently slows down and stops while the earth is overtaking it. It is exactly the same illusion which you notice when you are riding on a railroad train which overtakes a slower train travelling on a parallel track. The slower train seems to be moving backward while yours is passing it.

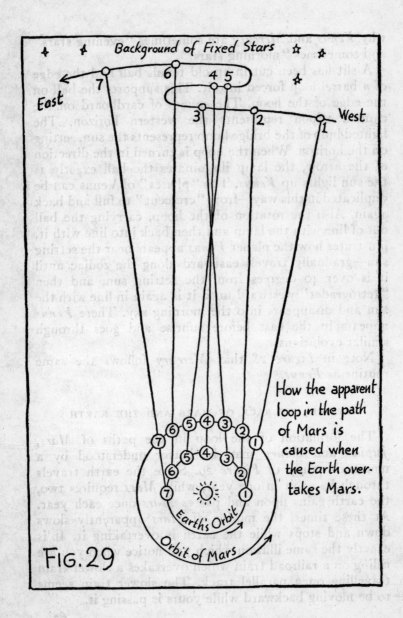

Background of Fixed Stars

East

West

How the apparent loop in the path of Mars is caused when the Earth overtakes Mars.

Earth's Orbit

Orbit of Mars

FIG. 29

The illustration in *Figure 30* shows how the hoop, tennis ball and lamp can also be used to illustrate the formation of the phases of the moon. As the observer turns with the hoop and ball, as the arrow indicates, he will see the moon go through all its phases, as shown by the small diagrams above.

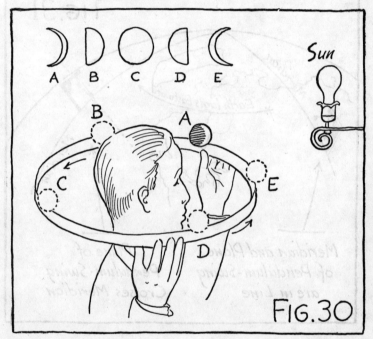

FIG.30

It is interesting to realize that, even after the Copernican explanation of the solar system's motions had been generally accepted, it still remained a theory. The actual proof of the rotation of the earth on its axis was not made until the famous "Foucault pendulum experiment" was devised by the French astronomer of the same name. (See *Figure 31*.)

In this experiment (first performed in the Pantheon in Paris) a heavy weight is set swinging in, let us say, the plane of the meridian. Since the tendency of the pendulum is to remain swinging in this original plane,

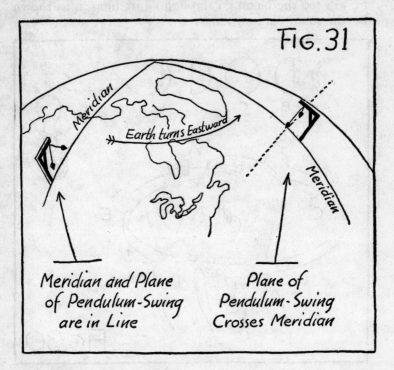

FIG. 31

Earth turns Eastward

Meridian

Meridian

Meridian and Plane
of Pendulum-Swing
are in Line

Plane of
Pendulum-Swing
Crosses Meridian

the earth's rotation soon rotates the point of the pendulum's support and the floor under it (in the original experiment, the roof and floor of the Pantheon) and the pendulum seems to alter the direction of its swing. This experiment enables us actually to view the rotation of the earth—through its apparent effect on the Foucault pendulum.

Incidentally, in computing the range for large caliber artillery, the motion of the earth must be figured in, so that the target may be hit successfully. The earth actually turns under the flying projectile (just as it does under the swinging pendulum). So in aiming a gun, we must make allowance for the distance the earth turns while the projectile is in flight.

SHOOTING STARS

On a clear night look at one part of the sky steadily for awhile. You will soon see a "shooting star" streaking across the heavens. Although people call it "a shooting star" it is no star at all. It is a *meteor,* or a small body, that flies through interplanetary space, and crashes into the earth's atmosphere. The friction of its movement through the air at high speed causes it to become white hot and burn. That's why it flares up and looks like a "falling star". Most meteors are completely burned up in the air. A few that manage to reach the ground are called *meteorites.*

METEOR SHOWERS

At certain times of the year there are showers of meteors coming from the same part of the sky. Each year, on August 12th, the *Perseids* come from the direction of *Perseus.* On November 14th the *Leonids* come from the direction of *Leo.* These showers are caused by swarms of rock that follow regular orbits around the sun. They occur at the same time each year when the earth crosses their orbit. These meteors are like little planets travelling around the sun. They are called *planetoids.* *Figure 52* on page 112 is a diagram of a Meteoric shower that can be seen near the constellation *Leo* every year about November 13.

THE STARS TEACH US TO GUIDE OUR SHIPS

TO ANY one who regards Astronomy as a rather impractical and highbrow hobby, its use in navigation is a sufficient answer. The safe arrival at their destinations of both ships and airplanes is of course entirely dependent upon accurate observations of the stars made with the instrument called the "sextant."

NAVIGATION DEPENDS UPON MEASUREMENT OF ANGLES

With a well-made sextant, in experienced hands, the latitude of a ship can be determined within a mile or two by measuring with the sextant the angular height of the heavenly body above the horizon.

This angle, plus the information given by the ship's compass and chronometer, enables the captain to hold a true course, and, when necessary, to go to any position radioed by a ship in distress.

Since finding the latitude of a ship or airplane is one of the chief applications of astronomy to human affairs, it is interesting for the star-gazing hobbyist to know a little about the principle involved.

To get a clear idea of it, we need only a few simple objects—including our trusty old umbrella with the stars chalked upon its inner surface. We shall also use the old tennis ball with a slit cut in it, which represented *Venus* and the *Moon* in the last chapter.

The split ball is placed around the umbrella rod, and a rubber band is put around its center, with the double purpose of holding the ball in place and of representing the equatorial line of the earth. (See *Figure 32*.)

HOW POLE STAR HELPS TO FIND LATITUDE

We are about to discover for ourselves how the *Pole Star* is used in finding latitude, since that is slightly easier than finding out how the sun is "shot" for the same purpose.

The next step is to cut out a disc of cardboard about an inch in diameter and put a pin through its center. The pin is then used to attach the disc to the tennis ball exactly half-way between the umbrella rod and the rubber-band equator. In other words, the pin indicates

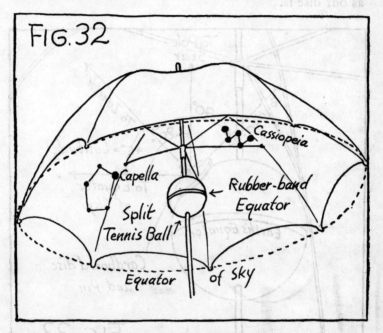

FIG. 32

Capella

Cassiopeia

Split Tennis Ball

Rubber-band Equator

Equator   of Sky

a point on the earth's surface at 45 degrees north latitude. (See *Figure 33*.) We know that the pin is at 45 degrees north because it is half-way between the equator (0 degrees) and the Pole (90 degrees North). But how will we locate our position if we are on a ship at this point on the earth, from which we can see no farther than the horizon circle, which is only about 20 miles from our ship?

69

It is now evident that the cardboard disc represents the approximately flat surface of the sea within a forty-mile circle. Since the earth's curvature is only a few inches to the mile, the comparatively small circle of sea visible from our ship can be regarded as practically flat, as our disc is.

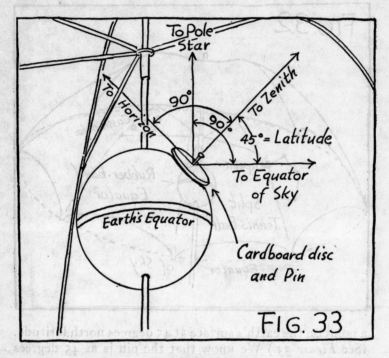

FIG. 33

In comparison with the millions and millions of miles separating the earth from the *Pole Star* and the equatorial belt of stars, the globe is a mere grain of sand. Accordingly, a line drawn from any point on the earth's surface parallel with the axis will point just as close to the *Pole Star* as the axis does. We are assuming of course, for our rough and ready purpose, that the axis does point exactly to the *Pole Star*.

AN IMPORTANT POINT TO UNDERSTAND

And, also because of the earth's infinitesimal size as compared with the distances to the equatorial stars, a line drawn from any point on earth parallel to the plane of the earth's equator will touch the starry globe in approximately the same plane as the equatorial plane does. When you have visualized the earth as a tiny grain of sand in interstellar space, you will have no further difficulty in understanding how latitude is found.

It then becomes obvious, as you will see from examining *Figure 33* for a moment, that the line of sight from an observer at the center of the disc to the *Pole Star* makes a right angle with his line of sight to the equator of the sky. It also becomes plain that another right angle is formed between his overhead point, or zenith, and the horizon of his place of observation.

NOW PLAY WITH TWO PAPER QUADRANTS

When you have verified these two angles, prepare two small quarter circles of writing paper. Letter them as shown in *Figure 34*, and hold them up against the light between the finger and thumb, with the points of the right angles coinciding.

As you cause these two little paper quadrants to slide upon each other around their common center, you will notice something very interesting. The line to the *Pole Star* on one quarter circle, and the line to the horizon on the other, will make an angle which is always exactly equal to the angle formed by the line to the zenith and the line to the equator. As one angle increases, the other increases with it, and vice versa.

Now for the way in which this interesting fact enables a ship captain to find his latitude from the *Pole Star*. Since these angles are always equal, the navigating officer has only to measure the angular altitude of the *Pole Star* above the sea's horizon in order to have the

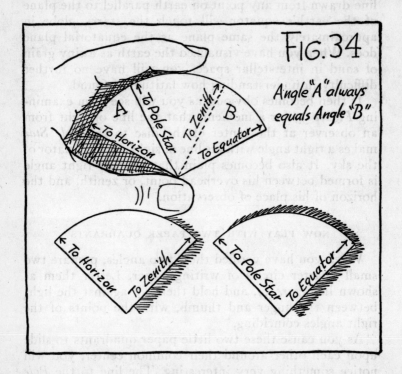

FIG. 34

Angle "A" always equals Angle "B"

angle between his zenith and the equator of the sky. And since the equator of the sky is in the same plane with the earth's equator, he has at the same time determined the angular distance of his point of observation to the equator—or, in other words, his latitude.

72

Take an example. The pin is inserted into the tennis ball halfway between pole and equator. An observer there would find the *Pole Star* 45 degrees above the horizon—and since this angle always equals the angle between his overhead point and the equatorial plane, he would at once know that his position is 45 degrees north latitude.

Finding the latitude by observing the altitude of the sun above the horizon at noon would be similar to finding it by the *Pole Star* if the sun travelled in the plane of the equator. In this case the latitude would be the angle between the equator and the zenith.

BUT SUNS DISTANCE FROM EQUATOR VARIES

But unfortunately it does not; it follows the line of the "ecliptic," which swings north of the equator in summer and south of it in winter. Accordingly, the skipper observes the sun's altitude at noon, and subtracts it from 90 degrees. He then adds to or subtracts from the remainder the angular distance between the sun and the equatorial plane on that day. This amount is given with extreme accuracy in the nautical almanacs published by all maritime countries—and is called the "sun's declination."

For example, a ship is lying on the 30th parallel of north latitude, as shown in *Figure 35*. The day is October 20, when the sun's declination is 10 degrees south of the equator. The captain's observation of the sun at noon shows it 50 degrees above the horizon. Fifty degrees subtracted from 90 leaves a remainder of 40 degrees. And since the sun is 10 degrees south of the equator on October 20, 10 must be subtracted from 40, leaving 30 degrees. This, being the angular distance between the zenith and the equator, is also the angular distance of the *Pole Star* above the horizon. And this figure, as we already know, is the latitude—or 30 degrees north.

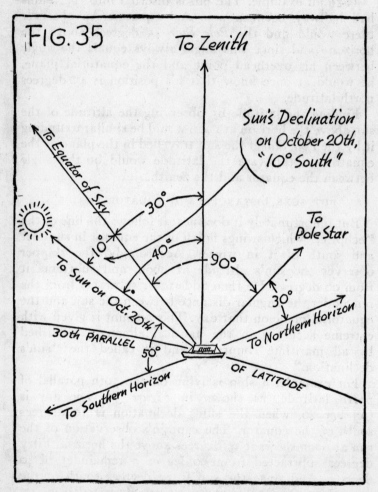

FIG. 35

To Zenith

Sun's Declination
on October 20th,
"10° South"

To Equator of Sky

30°

10°

40°

90°

To Pole Star

To Sun on Oct. 20th

30°

30th PARALLEL

50°

To Northern Horizon

OF LATITUDE

To Southern Horizon

THIS EXPLAINS THE PRINCIPLE

This brief outline of the latitude-finding process of course disregards the fine points which must be taken into account in actual navigation. One of these is the fact that the *Pole Star* is not exactly on the extended polar axis of the earth, but about two degrees distant from it. Our aim has been merely to make the principle plain.

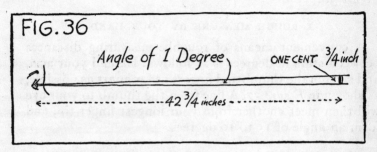

FIG. 36

Angle of 1 Degree

ONE CENT ¾ inch

42 ¾ inches

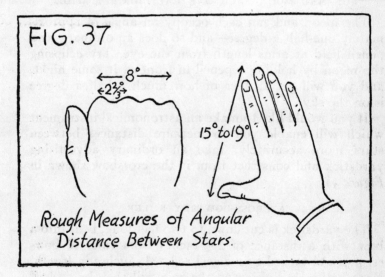

FIG. 37

8°

2⅔°

15° to 19°

Rough Measures of Angular Distance Between Stars.

HOW LONG IS ONE DEGREE?

It may, in this connection, be interesting to know just how much a degree is on the curve of the sky. Here is a simple rule:

Any round object, such as a copper cent, held at 57 times its diameter from the eye, will occupy an angle of one degree. The penny is ¾ inch in diameter; it must therefore be held 42¾ inches from the eye, as shown in *Figure 36*.

A ROUGH MEASURE BY YOUR HAND

A convenient means of roughly measuring distances between stars in degrees is as follows: Extend your arm, and stretch the thumb and fingers as far apart as possible, as shown in *Figure 37*. A line from the thumb to your eye will then meet another from your longest finger tip, and form an angle of 15 to 19 degrees.

SUN AND MOON EACH ONE HALF DEGREE WIDE

The moon and sun each occupy an angle of approximately one-half a degree—and so does an ordinary lead pencil held at arms length from the eye. Try eclipsing the moon by holding a pencil in front of it some night, and you will get an idea of how much a half a degree is on the sky.

If you would like to make an astronomical instrument which will enable you to measure distances between stars more accurately, take an ordinary advertising yardstick and construct from it the crossbow shown in *Figure 38*.

A CROSS-BOW SKY RULER

The yardstick is cut down to two feet long, bent into a bow with a distance of 2½ inches from ruler to bowstring, and tacked to a dowel rod 28½ inches in length. Each half-inch division on the rule will then be 57 half inches from the eye when the end of the rod is held on the cheek bone as shown. Accordingly, every half-inch on the rule measures one degree on the sky seen behind the rule. When you are out on a dark night a flashlight will enable you to see the half-inch divisions and at the same time see the stars between which you are measuring the distance.

ANCESTOR OF THE SEXTANT

Incidentally, this simple experiment enables you to rediscover, or re-invent, one of the earliest astronomical

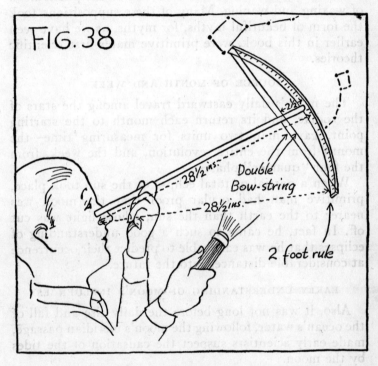

FIG. 38

28½ ins.

28½ ins.

Double Bow-string

2 foot rule

instruments used by mariners—the cross-staff. It enabled sailors of Columbus' time, and before, to measure the angular elevation of stars above the horizon—which is all a modern sextant does. (Directions for making a home-made sextant, a transit instrument, and other simple measuring devices, are given in Chapter IX.)

VII

WE DETECT ALL THE MOON'S MAGIC TRICKS

THE moon, because it is the nearest of all heavenly bodies to the earth, was the subject of a great many early speculations by man. He theorized about the spots on its surface as well as the phenomena of waxing and waning. Many of these suppositions took the form of beautiful myths, for myths, as we have seen earlier in this book, were primitive man's first scientific theories.

SOURCE OF MONTH AND WEEK

The moon's daily eastward travel among the stars of the zodiac, and its return each month to the starting point, gave man two units for measuring time—the month from the entire revolution, and the week, from the four "quarter" phases.

When a partial or total eclipse of the sun took place, primitive man had ocular proof that the moon was nearer to the earth than the sun, whose light was cut off. In fact, he came to such a good understanding of eclipses that he was early able to predict their occurrence at considerable distances into the future.

EARLY UNDERSTANDING OF MOON'S INFLUENCES

Also, it was not long before the daily rise and fall of the ocean's water, following the moon's meridian passage, made early scientists suspect the causation of the tides by the moon.

And finally the curving shadow of the earth, cast upon the moon during an eclipse of the latter, was probably a strong argument in favor of the earth's spherical form. All in all, man quite soon came to a fairly good understanding of the principal phenomena in which his nearest sky-neighbor took part.

One of the first things man must have noticed is that the moon is apparently considerably larger when it has just risen above the horizon than it is when sailing high in the sky. The fact that this is really only a persuasive optical illusion can be entertainingly demonstrated by a simple experiment. The wash-tub size of the moon as it appears above the eastern horizon is explained by this simple optical illusion. (See *Figure 39*.)

Take two strips of paper and cover everything above and below the line of dots from A to B. Then decide which distance seems longer—A to B, or B to C, judging entirely by the eye, using nothing with which to measure the relative distances.

WHY THE EYE AND JUDGMENT ARE FOOLED

They are actually equal; measure them and see. The line from A to B seems to the eye to be longer because it is filled with dots, while the line from B to C crosses empty paper. The distance between two points always seems longer to the eye if there are many intervening objects.

For the same reason, the moon on the horizon seems to be more distant than when it is overhead because the eye, in looking toward the horizon, takes in so many things on the way—trees, houses, hills, and so on.

The mind knows, of course, that the moon is really the same size whether in the zenith or just rising. But in spite of this, the illusion of greater distance to the horizon is so convincing to the eye that it seems the moon must be larger in order to appear like the same object it sees at the top of the sky, and as a result the brain back of the eye actually sees it larger.

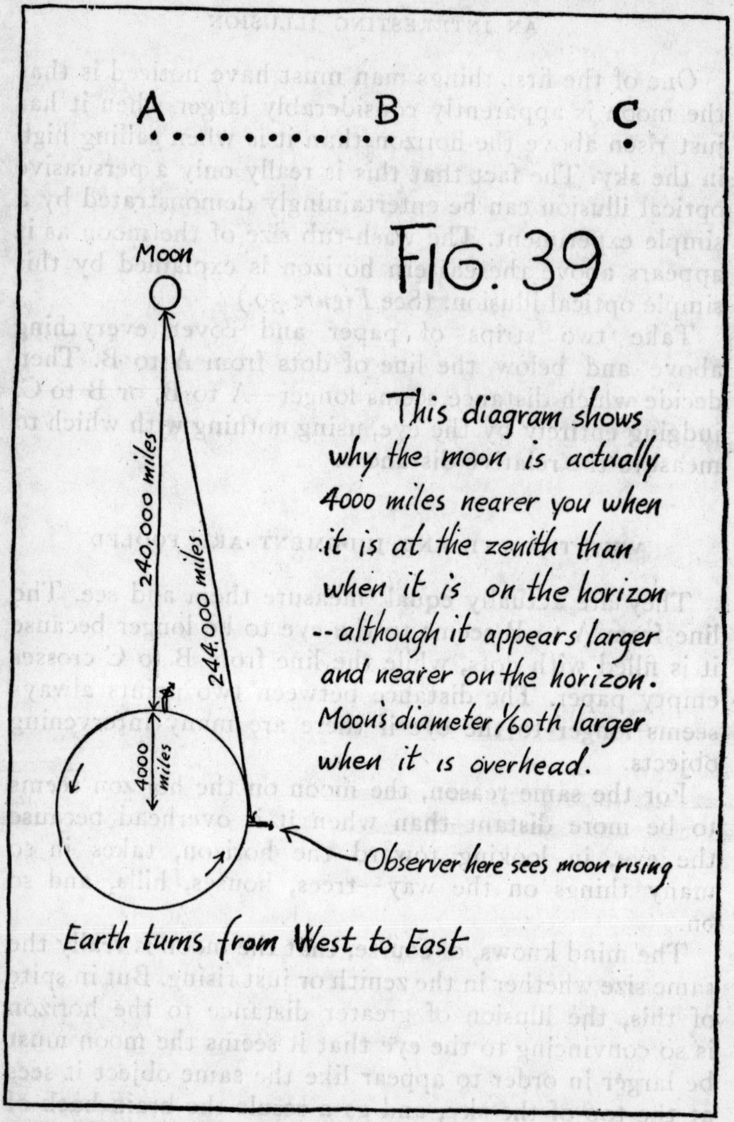

A · · · · · · · · · · B · · · · · · · · · · C

Moon

FIG. 39

This diagram shows
why the moon is actually
4000 miles nearer you when
it is at the zenith than
when it is on the horizon
-- although it appears larger
and nearer on the horizon.
Moon's diameter 1/60th larger
when it is overhead.

240,000 miles

244,000 miles

4000 miles

Observer here sees moon rising

Earth turns from West to East

To prove that the eye fools the mind in this way, roll up a narrow tube of paper about ten inches long and look through it at the washtub moon. It will instantly shrink to the size you see it when it is sailing overhead, for the tube cuts off the trees and hills that cause the illusion.

ACTUALLY, THE FACTS ARE OPPOSITE

The astronomical facts, shown in the diagram in *Figure 39* prove that the moon should really appear one-sixtieth larger when straight overhead, because the observer is then nearer to it by 4,000 miles, which is half the earth's diameter and one-sixtieth the distance to the moon.

In connection with the apparent size of the moon as effected by an illusion, you may find it interesting to measure the actual size of our satellite. You can perform this feat without instruments. This is the way to proceed· See *Figure 40*.

HOW TO MEASURE THE MOON

On a clear evening shortly before full moon, select a window through which the moon can be seen an hour or two after it rises. Then stick two strips of adhesive tape horizontally upon the glass about an inch and a quarter apart.

If you have a window with old-fashioned movable slat shutters, you can close one and use the space between the two adjacent slats instead of the tapes on the window pane. Finally, place a pile of several books on the corner of a movable table and you are ready for the observation.

The idea is to look at the moon between the tapes, move backward until the moon's image just exactly fills the space between them, and then measure accurately the distance of your eye from the tapes. The corner of the top book on the pile, brought close to the eye, will help in fixing the point where the eye-ball was when the moon exactly filled the space.

When this point is ascertained, measure carefully the exact distance from the book-corner where your eye was

Strips of adhesive tape

Measuring the Moon's diameter without instruments.

FIG. 40

to the tapes on the pane. Also verify painstakingly the exact width of the space between the tapes. Then you can calculate the diameter of the moon by a simple proportion in arithmetic.

The only figure you must take on trust is the moon's distance from the earth, roughly 239,000 miles. The other measurements needed you have made yourself. Here is how to figure out the moon's diameter:

DO IT BY SIMPLE PROPORTION

(The distance of tapes from eye) is to (the distance between tapes) as 239,000 is to (the moon's diameter).

It will be well to make your observation and measurements three times, taking the average of the three distances from the tapes to the eye.

Here are the results from the experiment as I tried it. Average distance of tapes to eye, 137.5 inches. Distance between tapes, 1.25 inches.

$$1.25 \times 239,000 = 298,750$$
$$298,750 \div by\ 137.5 = 2,170\ miles$$

The moon's diameter, measured in this crude way, comes out only a little larger than astronomers make it by the most refined methods.

HOW TO BE A TRUE PROPHET

Now that you have measured the moon's diameter, you can have fun predicting its position night after night among the stars. It will give you a thrill to see your predictions come true. Calculating the moon's position is very easy to do, provided you have made the "degree-measurer," or cross-staff, described in the chapter on the astronomy of navigation (Chapter VI).

You should begin on an evening when the moon is a narrow crescent in the western sky after sunset. Take your cross-staff degree measurer and place its right end at the moon. Let the rule extend to the left in a direction at right angles to a short line joining the horns of the moon's crescent. The rule will then lie along the moon's path among the stars.

Now read off 13 degrees on the rule, starting at the end on the moon. If there is a bright star near the 13 mark on the rule, note it carefully, for the moon will be near it on the next evening. If you measure off 13 degrees more to the left along the rule, it will mark the position where the moon will be on the second following night at the same hour.

In this way you can have some good fun making a map of the moon's future travels among the stars!

Begin by marking down the crescent moon near the right edge of a large sheet of wrapping paper. Make the crescent ¼ inch from horn to horn. Then rule a long line through the center of the moon's crescent, at right angles to a line joining the horns. This, as you know, is the direction of the moon's path, or "ecliptic."

PLOT THE MOON'S COURSE AHEAD OF TIME

Now note the bright star which is nearest to the moon and its path in the sky. Measure its distance from the moon on the sky in "degrees." Then mark its position and distance from the "moon" on your sheet of paper. Then find the first bright star toward the east (the left) along the moon's path. Measure its distance from the moon in "degrees" and mark it down on your wrapping paper at the right distance, in half-inches, from the "moon."

Then measure off on your paper the positions of the moon for two or three nights in advance. Put the moon's daily positions down at intervals of 13 half-inches (6½ inches).

On successive evenings you will then get quite a thrill at seeing how closely you have predicted the moon's travels among the stars.

You will probably wonder why I told you to draw the moon ¼ inch across on your map. This is because the moon occupies an angle of ½ a degree in the sky (or ¼ inch on the paper). The moon accordingly travels 26 times its own breadth in 24 hours, or about its width in an hour.

AN OCCULTATION

Some evening, when the moon is very near a bright star, you may see the east edge of the moon pass in front of the star. It will be fun then to get out your telescope and watch for the star's reappearance about an hour later on the west side of the moon. This obscuring of a star by the moon is called an "occultation."

Since the moon is about 2000 miles in diameter, and travels its own breadth in an hour, you can form some idea of how fast it whizzes through space!

WHAT IS THE SHAPE OF MOON'S PATH IN SPACE?

And what is the shape of the orbit in which the moon travels? Since it goes around the earth once every month, we should probably most of us try to draw the moon's orbit with a succession of loops or "epicycles." (See Chapter V.) But this is wrong, because the moon's motion with the earth along the earth's orbit is fast enough to prevent the moon from ever going backwards along the orbit.

MOON'S PATH AS SEEN FROM THE SUN

If our world were stationary in space, the moon would travel around it in a nearly circular path. But the earth and the moon are both revolving around the sun, which makes the actual path of the moon a slightly wobbly circle with the sun as its center. The moon weaves somewhat tipsily back and forth across the earth's

85

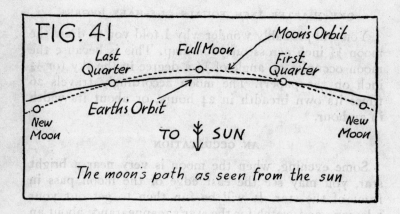

FIG. 41

Moon's Orbit

Full Moon

Last
Quarter

First
Quarter

Earth's Orbit

New
Moon

New
Moon

TO ☿ SUN

The moon's path as seen from the sun

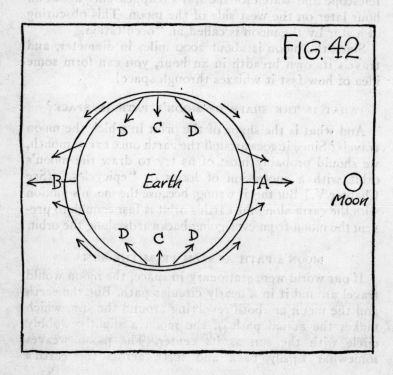

FIG. 42

D   C   D

B   Earth   A

D   C   D

Moon

path, spending half a month at a time on each side. If the earth's orbit is drawn on the ground with a diameter of 33 feet, the moon's path will not deviate from it by more than an inch.

HOW TIDES ARE CAUSED

The sketch in *Figure 42* shows that the moon raises two tidal waves in the earth's oceans. It also shows why this is so, and represents, by arrows, the directions of the forces which cause the earth's tidal bulges.

The ocean at *A* is nearer to the moon than the solid earth. Its strong attraction toward the moon causes the ocean surface to pull away from the earth. On the opposite side of the earth, at *B,* the ocean is further from the moon than the solid earth. The solid earth's attraction toward the moon is stronger than the pull on this part of the ocean. At both points the tendency to separate the ocean from the earth causes a bulge in the water, known as high tide. At points *C* the moon's attraction tends to draw earth and ocean together. This is low tide. At the intermediate positions the arrows show the direction of the force that results from the combined attractions on earth and ocean. Since the earth is rotating, different parts of its surface keep moving into the high tide positions *A* and *B*. That is why the high tide bulges travel around the earth.

ECLIPSES OF THE MOON

Anything that stops sunlight casts a shadow. The earth casts a shadow that points away from the sun. As the earth rotates, most of its surface spends part of the day in the sunlight and part of the day in the earth's shadow.

The moon does not shine by its own light. When it is out in the sunlight, it reflects some of this light back to us. But, as the moon travels around the earth, it sometimes enters the earth's shadow, where it receives no direct rays of sunlight. When this happens there is an eclipse of the moon.

The mechanism of eclipses of the sun is so simple that it is understood by every one. The moon simply comes between the earth and the sun—and, when all three bodies are exactly in line—a total or partial eclipse occurs. A total eclipse occurs only when the entire disc of the sun is totally obscured by that of the moon.

But did you ever stop to think what an amazing coincidence it is that the discs of sun and moon are almost exactly the same diameter?

SUN AND MOON OCCUPY SAME SIZED ANGLE!

If it were not for this strange coincidence of size—if the moon were a little smaller—or at a different distance from the earth—nobody would ever have witnessed the famous and awesome phenomenon we call a total eclipse of the sun!

This is how it becomes possible:

The moon is roughly 2000 miles in diameter; the sun is approximately 400 times as wide, or over 800,000 miles. The moon is about 240,000 miles from the earth. If the sun was exactly 400 times farther, it would be about 96,000,000 miles. As a matter of fact, it is around 93,000,000 miles.

Is it not a really surprising coincidence of distances and sizes, causing both sun and moon to occupy the same angle in the eye of an observer on earth?

One more of the moon's magic tricks and we must leave the fascinating subject. We shall take it up again briefly in the next chapter in considering opera and field glass observations of our satellite.

"HARVEST" AND "HUNTER'S" MOONS

This last trick of the moon is truly magical, for it produces the wonderful Autumn full moons which are called "Harvest Moon" and "Hunter's Moon."

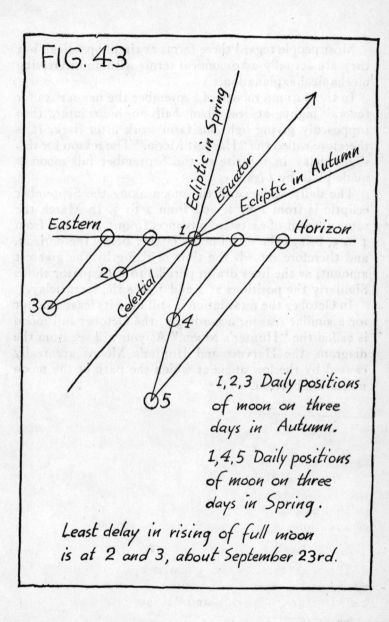

FIG. 43

Ecliptic in Spring

Equator

Ecliptic in Autumn

Eastern

Horizon

1

2

3

Celestial

4

5

1, 2, 3 Daily positions
of moon on three
days in Autumn.

1, 4, 5 Daily positions
of moon on three
days in Spring.

Least delay in rising of full moon
is at 2 and 3, about September 23rd.

Most people regard these terms as simply poetical, but they are actually astronomical terms with an interesting mechanical explanation.

In the autumn month of September the moon rises for several nights at less than half an hour later, thus supposedly giving light for farm work after dark. It is therefore called the "Harvest Moon." The reason for this slight delay in the rise of the September full moon is made plain by *Figure 43*.

The daily advance of the moon along the September ecliptic is from 1 to 2, and from 2 to 3. In March the same amount of eastward advance, from 1 to 4, and from 4 to 5, brings the moon much farther below the horizon, and therefore retards the time of rising by the greatest amount, as the lines drawn parallel to the equator show. Similarly the positions at 2 and 3 give the least delay.

In October the retardation is still near its least amount for a similar reason; accordingly, the October full moon is called the "Hunter's Moon." As you will see from the diagram, the Harvest and Hunter's Moons are really caused by the low angle at which the path of the moon meets the horizon.

VIII

WE FOLLOW GALILEO'S TRAIL WITH OUR OPERA AND FIELD GLASSES

SINCE many of the finest achievements of modern astronomy have been made possible by the great enlarging powers of telescopes, it will be interesting to rediscover just how much could be seen with the first astronomical telescope—the one that the Italian scientist Galileo made for himself. This is one case where we can actually have the thrill of seeing what a pioneer discovered for the first time.

THE FIRST TELESCOPE OF ALL

Although the telescope was invented in Holland, about 1608, by Hans Lippersheim, a spectacle maker, the one which Galileo built was the first ever used for astronomical discovery.

When Galileo heard of the Dutch invention, he had no detailed information about the telescope's construction, but a little thing like that never stopped this resourceful investigator. The prospect of having an instrument which would magnify the heavenly bodies spurred his inventiveness. Accordingly, after a few trials, Galileo succeeded in arranging two lenses (one convex and one concave) in a tube in such a way as to enlarge the size of the distant object looked at. Galileo was therefore the inventor of the opera glass, which is merely a short pair of telescopes built from these two kinds of lenses.

ABOUT AS STRONG AS AN OPERA GLASS

The first telescope Galileo constructed was, in fact, no stronger than the average opera glass, which magnifies about three diameters. However, he kept on experimenting with grinding and combining lenses of different powers until he was able to produce a telescope magnifying 30 diameters.

HOW TO FIND THE POWER OF A GLASS

Perhaps, having mentioned the subject of telescopic magnification, it will be advisable to take a short detour at once and explain how to determine the magnifying power of any glass you may have to deal with.

Point your opera glass, field glass or telescope at the side of a house. Have the house just far enough away so that you can count the rows of bricks (or the white clapboards) with your unassisted eye.

Then focus your glass upon the clapboards (or bricks) of the housewall. When the image is sharp, look with one eye through the instrument at the clapboards, and with the other naked eye at the house. One eye will then see the magnified image of the clapboards or bricks through the glass, and right beside the magnified bricks, your other eye will see them in their natural size.

HOW MANY NATURAL SIZED ONES TO ONE MAGNIFIED ONE?

After a moment's practise, you can see the little bricks and the big bricks at the same time, side by side. When you have succeeded in doing this, count how many of the little bricks (seen by your unassisted eye) are equal in height to one big brick (seen by your other eye through the glass). If you find that it takes about 6 of them, the glass you are using is "6 power," or magnifies 6 diameters.

GALILEO BEGAN WITH THE MOON

And now to return to Galileo and the 30 power telescope which he finally succeeded in building. It was natural for him to turn his newly-made instrument first

92

upon the moon, since it is the most prominent heavenly body.

And no sooner had Galileo taken a good look at the magnified moon than he knew he had made a remarkable discovery.

Previous to the invention of the telescope, it was commonly believed that the moon, like all the other celestial bodies, was perfectly smooth and spherical. The cause of the dark markings which outline the man in the moon, or the lady, was quite unknown.

"THE MOON HAS MOUNTAINS! FOUR MILES HIGH!"

Galileo discovered at once that many of the smaller dark markings were the shadows of lunar mountains cast by the sun. As he watched the moon wane on successive nights through his telescope, he could see the shadows of some particular peak lengthen and the entire mountain eventually go into the darkened part of the moon. Galileo even made a guess at the probable height of some of the more conspicuous mountains. The largest was estimated by him to be about four miles high, a result which agrees surprisingly well with our modern measurements.

The method adopted by Galileo of studying the mountains of the moon when they are on the line where dark and light parts join is still the best way to observe them.

HOW TO OBSERVE MOON-FEATURES

When you start to observe and study the moon through your glass, do not make the mistake of choosing the time of full moon. Why not? Because at full moon the sunlight (which makes the moon shine by reflection) is illuminating the whole surface—just as the sun does the earth at noon.

93

You know that there are no large shadows cast when the sun is overhead, but that the shadows get longer and blacker in the morning or in late afternoon. It is the same on the moon. You can study it best when the shadows of the mountains are longer and stronger.

WHAT IS THE "TERMINATOR?"

The very best way to learn the principal features of the moon is to watch the place where the dark and light parts of the moon join. This line where light meets shadow is called the "terminator," and it advances night after night across the face of the moon as it waxes from thin crescent to full and then wanes from full to crescent again.

The three small pictures of the moon at different ages (See *Figure 44*) will help you to become familiar with a few of the chief features during the progress of a month.

After making this discovery proving the real nature of the moon's surface, Galileo literally looked for "more worlds to conquer" with his telescope.

GALILEO'S SECOND DISCOVERY

Bright *Jupiter* attracted him, and on January 7, 1610, he turned what he called his "optic tube" in the direction of the giant planet.

He was able to see the disc of the great world, and noticed near it three faint little stars which caught his attention on account of their nearness to the planet. It also struck him as remarkable that they were nearly in a straight line with *Jupiter*.

Next night Galileo looked at *Jupiter* and his tiny companions again, and noticed that they had changed their positions relative to the planet's disc. But the change did not seem to be the result of *Jupiter's* movement in his orbit, as it would have been if the tiny bodies were fixed stars. Two nights later, Galileo decided that

94

the bodies were moving around *Jupiter*, and when, a few days later, he found a fourth one (which had been hidden behind the planet) he had completed the discovery of all Jupiter's moons.

THE SATELLITES WITH OPERA OR FIELD GLASS

A strong opera glass will give you a hint of Jupiter's moons when they are not too close to the planet, and when there is no moonlight to dim them. With a field glass magnifying eight or more times, you will have no difficulty in seeing all of the moons, unless they are behind or in front of the planet, or so close to his disc that they are lost in his bright glare.

GET A NAUTICAL ALMANAC

If you wish to observe *Jupiter's* satellites in a serious way, send to the Government Printing Office, Washington, D. C. for a copy of the American Ephemeris and Nautical Almanac for the current year, price $3.75 clothbound. It gives the position of *Jupiter's* moons for every day in the year when the planet is in a convenient position for observation.

FIRST OBSERVATION OF SATURN

The second of the planets to which Galileo applied the powers of his telescope was *Saturn*. His instrument was, however, not strong enough to enable him to make the discovery of the famous ring. This was reserved for *Christian Huygens*, who made it over 40 years later with a considerably more powerful instrument. *Galileo* was only able to note with his that the planet *Saturn* seemed to be composed of three parts: a larger central one, and two smaller parts on each side.

DISCOVERY OF VENUS' CRESCENT

But when the great Italian looked upon the magnified *Venus*, he found out something well worth while. He

96

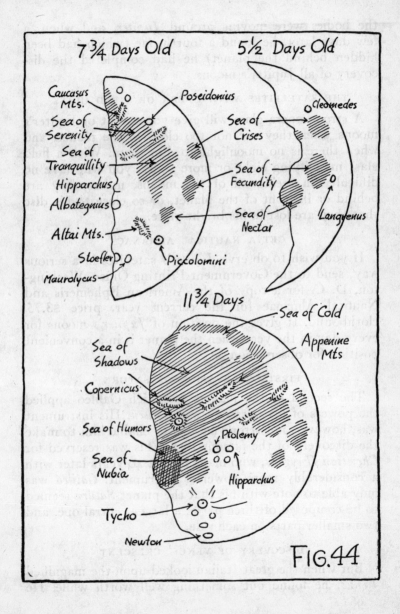

7¾ Days Old

Caucasus Mts.
Poseidonius
Sea of Serenity
Sea of Tranquillity
Hipparchus
Albategnius
Altai Mts.
Stoefler
Maurolycus
Piccolomini

5½ Days Old

Cleomedes
Sea of Crises
Sea of Fecundity
Langrenus
Sea of Nectar

11¾ Days

Sea of Cold
Appenine Mts
Sea of Shadows
Copernicus
Sea of Humors
Sea of Nubia
Ptolemy
Hipparchus
Tycho
Newton

FIG. 44

found that *Venus* passed through phases very similar to those of the moon. This was direct evidence against the Ptolemaic system.

AND FINALLY, SUNSPOTS

The last of the great contributions which Galileo's pioneer glass made to science was the discovery of the nature of the dark spots on the sun. These had of course been seen occasionally before with the naked eye, but had been believed to be caused by the planet Mercury passing across Old Sol's face. Galileo however studied them, made drawings of their shapes, proved that they rotated with the sun, and decided that they were of the nature of blemishes upon the solar surface.

Galileo seems not to have been very much interested in *Mars*, probably because it requires a stronger glass than he possessed to see the distinctive features such as the white polar caps and the so-called "seas."

But after Galileo had observed everything that he could about the planets and the moon, he never tired of observing the other features of the sky which fell within the scope of his instrument.

ASTRONOMY FASCINATING WITHOUT
POWERFUL GLASSES

Many people are kept from enjoying the pleasures of Astronomy by the belief that nothing worthwhile can be seen without a powerful telescope. That this is erroneous is proved not only by the experience of Galileo but also by that of many others. A good modern achromatic glass is far better than any instrument which the early observers had at their disposal. With even a good strong opera glass you can go upon a fascinating tour of discovery in the sky—following the trail of Galileo at every step.

97

The page of small circular detail maps (*Figure 45*) shows you a few of the most interesting features which you can observe with opera and field glass. To locate these features among the constellations, you will need

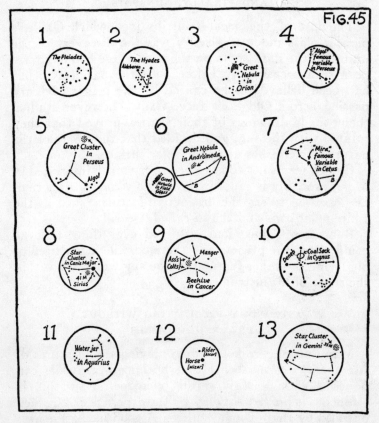

FIG. 45

1 The Pleiades
2 The Hyades Aldebaran
3 Great Nebula in Orion
4 Algol famous variable in Perseus
5 Great Cluster in Perseus Algol
6 Great Nebula in Andromeda Great Nebula in Field Glass
7 "Mira," famous Variable in Cetus
8 Star Cluster in Canis Major 41 M Sirius
9 Ass's Colts Manger Beehive in Cancer
10 Deneb Coal Sack in Cygnus
11 Water jar in Aquarius
12 Rider [Alcor] Horse [Mizar]
13 Star Cluster in Gemini

to turn back to *Figure 11*. In these maps of the star groups you will see small figures in circles. Each of these refers to one of the numbered circular maps in *Figure 45*. After locating the feature to be observed in the proper constellation, you can refer to *Figure 45*

for the details. The following numbered paragraphs describe the various features. They can of course be observed at other times of the year than those mentioned. Consult *Figure 11* for the times when the proper constellations are on the meridian or in view.

1. *The Pleiades*

This beautiful little group is high in the eastern sky during December, and equally high in the southern sky in January. It will require a powerful field glass to see all the stars represented in map 1, but an opera glass shows many more than the naked eye does, and reveals their rich beauty. *The Pleiades* were among the first objects studied by Galileo with his first telescope. He counted 36 stars in the group, which was beautifully described by the poet Tennyson in the phrase—"a swarm of fireflies tangled in a silver braid." *The Pleiades* are part of the constellation called *Taurus*, or "the Bull."

2. *The Hyades*

This group, also in *Taurus*, is clustered near the first magnitude star *Aldebaran*, which forms the *Bull's* eye. It is somewhat similar to the *Pleiades*. An opera glass gives you a beautiful picture—and a stronger field glass enables you to see many more faint stars, including several double stars.

3. *The Great Nebula in Orion*

This is only one of the many beauties exhibited by this magnificient star group, which hangs in the southeastern sky during December and January, and crosses to the southwest by March. No other constellation contains so many bright stars. Look for the *Nebula* with your glass about half way down the row of stars which hangs like a sword from the famous "belt." Even an opera glass will show you faint stars in the *Nebula* itself—and a field glass makes them much plainer.

4. *Algol—Famous Variable Star*

You will find *Algol* in the constellation *Perseus*—high up in the eastern sky in December; and near the zenith of the southern sky in January. *Algol*, or "the *Demon*," goes through its changes in brilliancy in about 9 hours. At its brightest it is a second magnitude star (like those in the *big dipper's* handle). It remains bright for 2½ days. Then it fades for 4½ hours—and becomes an inconspicuous 4th magnitude star—which it remains for a few minutes only. Its increase in brightness also takes 4½ hours—after which the cycle begins again. The interval between times of minimum brightness is 2 days, 20 hours and 49 minutes—so you can alculate its periods ahead for yourself after observing one of them.

5. *The Great Cluster in Perseus*

To see this lovely object we must look also in *Perseus*, shown also in map No. 4. The cluster can be seen with the naked eye as a fuzzy star at the top of the constellation. An opera glass brings out some of the many faint stars—a field glass more—and a small telescope makes it a magnificient ball of suns!

6. *The Great Nebula in Andromeda*

This famous object is located high in the western sky for December, when it should be observed, as by January it is on its way to the horizon. It can be seen with the naked eye on a moonless night. With an opera glass it shows as a wisp of light like that indicated in the small circle beside Map No. 6. The stronger the glass or telescope, the more beautiful and interesting the *Nebula* becomes.

7. *The Variable Star Mira, in Cetus*

Look for this star in the group called *Cetus*, or the *Whale*, about half way to the zenith in the southern sky for December. It fades from the second magnitude to about the tenth (when it is invisible to the naked eye for 5 months). Eleven months are required for this long-period variable to fade and come back to its full brightness. No wonder the ancients named it *Mira*, or "the wonderful one."

8. *Star Clusters in Canis Major*

Southeast from *Orion*, the blazing first magnitude star *Sirius* will guide you to a field that richly repays study with an opera or field glass. The cluster marked "41 M" in the map is only one of several in the near neighborhood or *Sirius*.

9. *Praesepe, or the "Beehive," in Cancer*

The faint constellation called *Cancer*, or "the *Crab*," is rising in the eastern sky during January, and about half way up the sky in February. At its center your opera glass will show you a crowd, or cluster, of lovely little stars, with a brighter one on each side. This cluster is also called "the *Manger*" and the two stars the "*Ass's Colts*," feeding from it. Galileo's telescope enabled him to count 36 stars in the *Manger*.

10. *The Coal Sack in Cygnus*

Here also the Milky Way is glorious in a glass, but is marked by a black spot, called the "coal sack." It may be caused by a cloud of dark nebulous matter.

11. *The Water Jar in Aquarius*

You will find this striking little triangle above the southwestern horizon in early December. Sweep your glass southward and you will see a lovely stream of

little sparklers, which represent the "water" being poured from the "jar."

12. The Horse and Rider

The star at the bend in the *Big Dipper's* handle is the "*Horse*" (*Mizar*). The "*Rider*" (*Alcor*) is immediately above it. The *Rider* was also called "the *Proof*" by the Arabs, meaning the proof of good eyesight if you can see it.

13. The Star Cluster in Gemini

In this group called *Gemini*, or "the *Twins*," look with your opera glass near its western end for the cluster "35 M" at the center of a gorgeous field of small stars. A field glass brings out more details in the cluster and its surroundings.

This baker's dozen of field glass features is not meant to be more than a hint. If you wish to go on and explore the sky more thoroughly, you will need a more complete guide than this book could be. There is nothing better than *Astronomy with an Opera Glass*, by Garrett Serviss. Other good books are *A Field Book of the Stars*, and *In Starland with a Three Inch Telescope*, both by William Tyler Olcott.

By the time you become the possessor of the last-named volume you will have graduated from opera and field glass astronomy and will undoubtedly be fired with the ambition to buy or build a telescope which will enable you to see more than even the strongest field glass or spy-glass will reveal.

Only those who actually arrive at some comprehension of the marvelous beauty and orderly system which pervade the universe can understand the continued fascination which the hobby of astronomy has for its devotees.

IX

PHOTOGRAPHS AND BOOKS

On the following pages are photographs and drawings of some of the more interesting stars, families of stars, and planets. There are also illustrations of well-known modern telescopes.

Not everyone gets the chance to look through a high-powered telescope, but everyone can own some of the very good photographs taken with them. If you want to start your own collection, you may obtain excellent pictures by writing to the following sources for descriptive literature:

Yerkes Observatory, Williams Bay, Wisconsin,

Mt. Wilson and Palomar Observatories, Pasadena, California.

The planetarium nearest you also has good source material available, from which you can choose.

If you want to bring the stars closer to you, a pair of powerful binoculars will be of considerable help. When you become more advanced, you may want to own your own telescope. If you are unable to purchase a good telescope, you may be interested in assembling one from ready-made parts that are sold as kits. These are advertised in any of the astronomy magazines listed on page 105. There may also be an amateur astronomy club near you that provides instructions and facilities for building your own scope. The names and addresses are often listed in the amateur astronomers' publications.

For those who want to know more about the science of astronomy, there are lots of exciting books to choose from. Following are a few of the many books that are available:

CREATION OF THE UNIVERSE. *by George Gamow.* Viking Press, New York. $3.75. Deals with the fundamental questions about the universe as a whole. Fully illustrated.

PICTORIAL ASTRONOMY. *by Dinsmore Atler and Clarence H. Cleminshaw.* Thomas Y. Crowell Co., New York. An excellent guide to a fascinating hobby. A large, lavishly illustrated book, full of charts, photographs and drawings.

THE NATURE OF THE UNIVERSE. *by Fred Hoyle.* Harper's, New York. $2.50. The larger problems of space and time are discussed in a clear and forceful manner.

THE HANDBOOK OF THE HEAVENS. *by Bernhard, Bennet and Rice.* McGraw Hill, New York. $4.50. Offers the beginner a lucid explanation of the fundamentals of astronomy. Written in a text book style.

INSIGHT INTO ASTRONOMY. *by Leo Mattersdorf.* Scott Publishing Co., Cambridge, Mass. $3.50. A simple, easy to understand presentation of the mysteries of the heavens.

HISTORY OF ASTRONOMY. *by Georgoi Abetti.* Schuman Inc., New York. $6.00. An excellent presentation of the development of astronomy, the oldest and the noblest of all sciences.

THE STARS. *by H. A. Rey.* Houghton-Miflin, Boston, Mass. $4.50. A new way to look at the stars. Outlines that can really be seen. Includes a large star chart and a section on basic astronomy.

STARS IN THEIR COURSE. *by Sir James Jeans.* Cambridge University Press, New York. $1.95. The layman's introduction to the science of astronomy.

SKYSHOOTING. *by Mayall and Mayall.* Ronald Press, New York. $3.75. Practical instructions for taking photographs of the heavens.

PRIMER FOR STAR-GAZERS. *by Harry M. Neely.* Harper's, New York. $5.00. This practical book helps the beginner to readily find any constellation or star in the heavens.

THE PLANET MARS. *by Gerard de Vaucouleurs.* MacMillen, New York. $2.50. Discusses the various facts and theories about Mars.

104

MAKING YOUR OWN TELESCOPE. *by Allen Thompson.* Scott Publishing Co., Cambridge, Mass. $4.00. Easy-to-follow, step-by-step instructions for building a reflecting telescope.

THE UNIVERSE WE LIVE IN. *by John Robinson.* Thomas Y. Crowell Co., New York. $4.50. An exciting account of our world and the world around us.

THE UNIVERSE OF DR. EINSTEIN. *by Lincoln Barnett.* $2.75. An interesting and authoritative discussion of the structure of the universe as it is presently known to science.

SUN, MOON, AND PLANETS. *by Roy K. Marshall.* Henry Holt and Co., New York. $2.50. An excellent introduction to the solar system.

THE WORLD OF COPERNICUS. *by Angus Armitage.* Mentor Books. 35¢.

THE SIZE OF THE UNIVERSE. *by F. J. Hargreaves.* Pelican. 35¢.

The following pamphlets are also available:

SPLENDORS OF THE SKY (third edition, revised). Scott Publishing Co., Cambridge, Mass. 50¢. Includes 36 pages of interesting photographs.

OBSERVER'S HANDBOOK. Available at most plantariums for 55¢.

PALOMAR OBSERVATORY. Available at most planetariums for 75¢.

You will also find the following periodicals helpful:

SKY AND TELESCOPE. A monthly magazine of the Sky Publishing Co., at Harvard College Observatory, Cambridge 38, Mass.

SKY REPORTER. Monthly bulletin of the Hayden Planetarium, at 81st Street and Central Park West, New York 24. 10¢ a copy.

MONTHLY EVENING SKY MAP. A quarterly, published by the Celestial Map Publishing Co., P. O. Box 3, Pike County, Shela, Pa. 60¢ a copy.

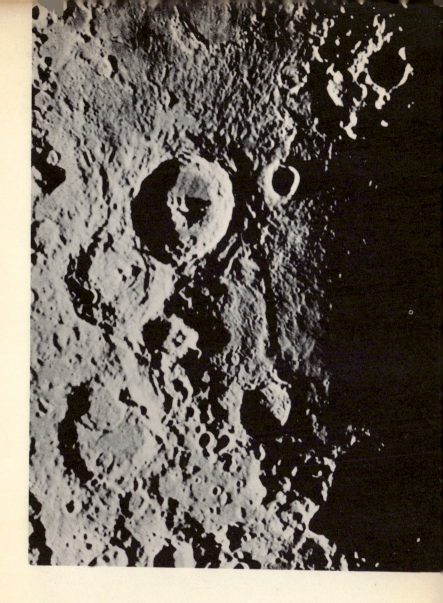

Fig. 46. Craters on the face of the moon, first seen by Galileo when he looked through his 30-power telescope (see page 92). *Courtesy of Yerkes Observatory.*

Fig. 47. Great Spiral Nebula of *Andromeda,* a universe, similar to our own Milky Way (see pages 96 and 100). Photographed through a 100-inch telescope.

Fig. 48. 25 foot Radio Telescope at Harvard Observatory. Can "see through" dust clouds, using the same principles as radar (see page 130).

Fig. 49 (*upper*). Globular Star Cluster in *Canes Venatici;* a constellation in the northern sky, between *Big Dipper* and *Boötes.* Photographed through 200 inch telescope.

Fig. 50 (*lower*). Interior view of 200 inch Hale Teelscope (see page 126). Drawn by Russel W. Porter.

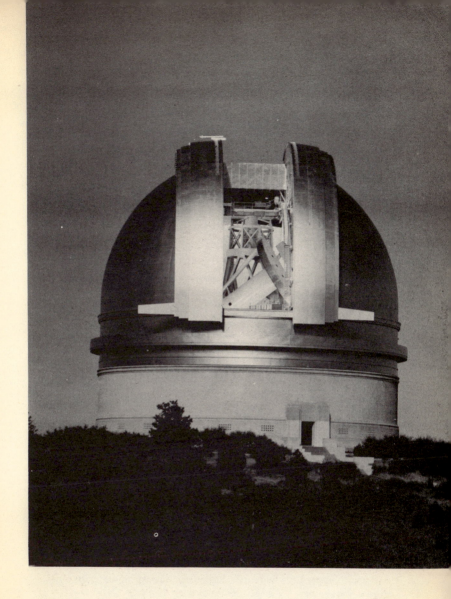

Fig. 51. Moonlight view of 200 inch Hale Telescope dome. The dome moves on tracks, and can be turned so that the telescope faces any part of the sky.

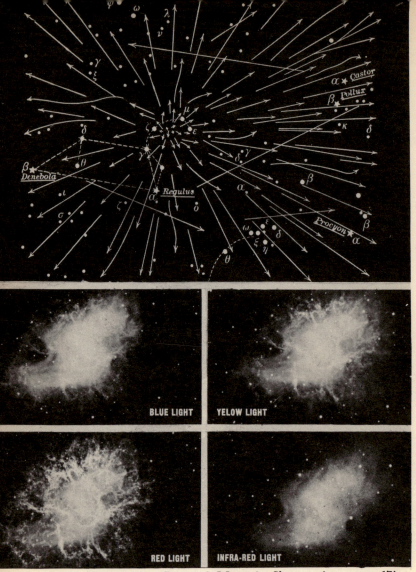

Fig. 52 (upper). Chart of Meteoric Shower (see page 67), drawn November 15, 1886. *Courtesy of Ginn & Co.*

Fig. 53 (lower). Crab Nebula in *Taurus.* 4 photos taken through 100 inch telescope with different color lights. Made up of glowing gases, possibly blown out of star by atomic explosion.

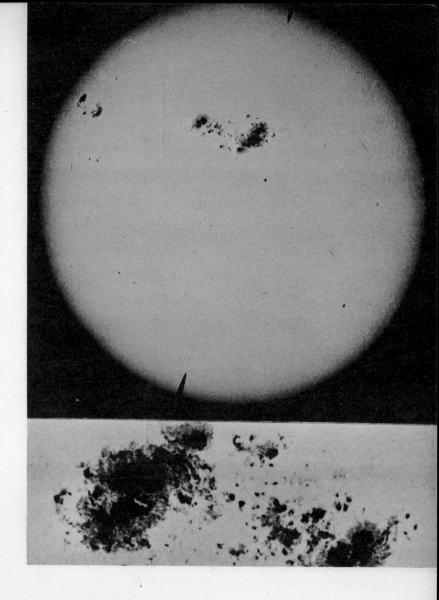

Fig. 54. Photo of the sun's face (upper), and enlargement of the Great Sun Spot of April 7, 1947 (lower). The process that creates the sun's energy is also used in the H-bomb (see page 55).

Fig. 55. Six types of extragalactic nebulae. These are "island universes" like our own Milky Way (see page 125).

SIMPLE INSTRUMENTS YOU CAN MAKE

YOU CAN get acquainted with some of the methods used to study the stars, by employing simple, home-made instruments. The measurements you make with them are rough but useful.

MEASURING BRIGHTNESS

A star's brightness is measured on a special scale of star "magnitudes" in which higher numbers indicate fainter stars. First magnitude stars are brighter than second magnitude stars, and so on. The constellation *Draco* provides an easy way of learning how to judge the brightness of a star . The head of *Draco* (See page 35) is made up of four stars that form an irregular quadrilateral. The brightest of these is a first magnitude star. The second brightest is a second magnitude star. The third brightest is third magnitude, and the faintest of the four is fourth magnitude. You can find out the magnitude of other stars by comparing them with these four stars in the head of *Draco*. You can see only up to the sixth magnitude with the naked eye. A telescope must be used to see fainter stars.

A HOME-MADE SUNDIAL

In the daytime you can use the sun to tell time with the help of a sun-dial. This is nothing more than a stick that casts a shadow. The shadow points out the time like the hour hand of a clock.

As the earth spins on its axis, the sun appears to make a big circle in the sky around the axis. If we point a stick in the direction of the earth's axis, and we catch its shadow on a flat surface parallel to the equator, this shadow will keep time with the passage of the sun across the sky. To use this fact for making a sundial, follow these directions:

First look up your latitude in an atlas. Then subtract
it from 90 degrees. Now, cut two wedge-shaped pieces
of wood with the angle of the wedge equal to the result.
For example, if your latitude is 40 degrees, you subtract
it from 90 degrees and make the wedges 50 degrees
(*Figure 57, A*). Now cut two rectangular boards about
12" by 18". Draw a line parallel to the long edge, 1"
from the long edge (*Figure 57, B*). Through the center
of this line, nail a 4" dowel stick at right angles to the
board. On one side of this line divide the space into 12
angles of 15 degrees each, with the aid of your protract-
tor (*Figure 57, C*). Label the lines from 6 A.M. to 6
P.M., as shown in the diagram. Now nail the two boards
to the wedges so that the boards touch at one edge with
the hour lines on the upper board pointing away from
the edge, as shown in the diagram. Your sun-dial is now
complete (*Figure 57, D*). Before it can be used, however,
it has to be set in the right direction. The bottom board
must be level. The edge where the two boards meet
must run east and west. To level the bottom board use
a carpenter's level. The east-west line is, of course, at
right angles to the north-south line. To locate north

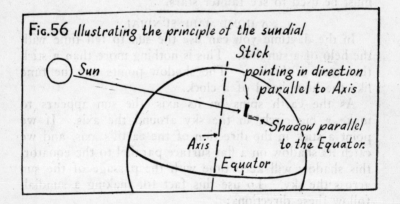

Fig. 56 *illustrating the principle of the sundial*

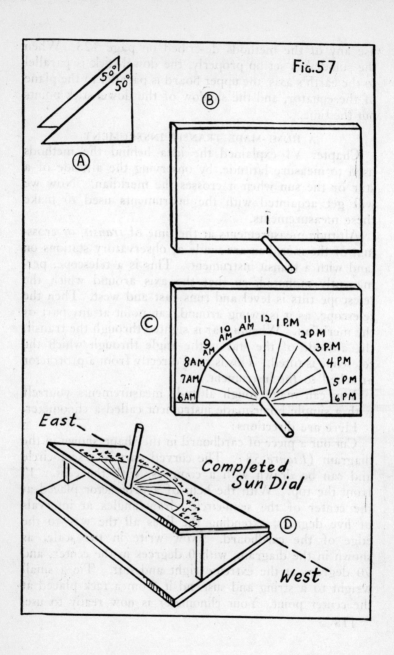

Fig. 57

50° 50°

B

C

12
11 1 P.M
10 2 PM
AM
9   3 P.M
AM
8AM 4 PM
7AM 5 PM
6AM 6 PM

East

Completed
Sun Dial

D

West

use any of the methods described on page 123. When the sun-dial is set up properly, the dowel-stick is parallel to the earth's axis, the upper board is parallel to the plane of the equator, and the shadow of the dowel-stick points out the time.

A HOME-MADE TRANSIT INSTRUMENT

Chapter VI explained the idea behind the methods used to measure latitude, by observing the altitude of a star or the sun when it crosses the meridian. Now we will get acqainted with the instruments used to make these measurements.

Altitude measurements at the time of *transit,* or crossing, of the meridian are made at observatory stations on land with a transit instrument. This is a telescope, permanently mounted, so that the axis around which the telescope tilts is level and runs east and west. Then the telescope, as it is swung around, can point at any part of the meridian. When a star is sighted through the transit, the altitude of the star is the angle through which the telescope is tilted. This is read directly from a protractor attached to the instrument.

You can make rough altitude measurements yourself with a simple home-made instrument called a clinometer. Here are directions:

Cut out a piece of cardboard in the shape shown in the diagram (*Figure* 58). The curved part is a semi-circle and can be drawn with a compass, with the center 1″ from the top. With the help of a protractor placed at the center of the semi-circle, draw angles at intervals of five degrees, extending the lines all the way to the edge of the cardboard. Now write in the scale, as shown in the diagram, with 0 degrees in the center, and 90 degrees at the extreme right and left. Tie a small weight to a string and suspend it from a tack placed at the center point. Your clinometer is now ready to use.

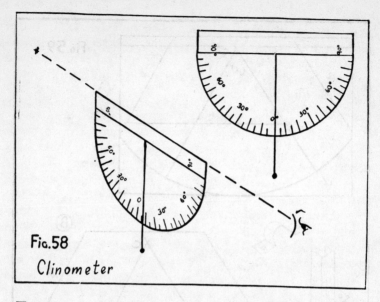

Fig. 58

Clinometer

To measure the altitude of a star, hold the clinometer up to the eye and sight the star along the straight edge of the cardboard. Keep the cardboard in a vertical plane. The string, held in a vertical position by the weight, shows the altitude of the star on the scale. To use the clinometer as a transit instrument, you must face north.

A HOME-MADE SEXTANT

When a transit instrument is used, you must be sure that its axis is really level. This is impossible on a pitching and tossing ship at sea. That is why a sextant is needed in its place. You can learn how a sextant works by making and using a simple, home-made one. On a piece of wood about a foot high and about 1½ feet wide, put a dot at *c*, about one inch below the top and half-way between the sides (See *Figure 59, A*). Using *c* as center, draw an arc of a circle, *d*, as shown. At *c* make an angle of 60 degrees with a protractor. From the points where

119

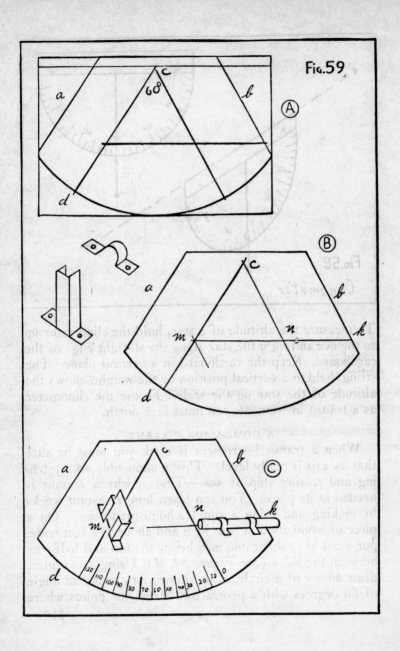

Fig. 59

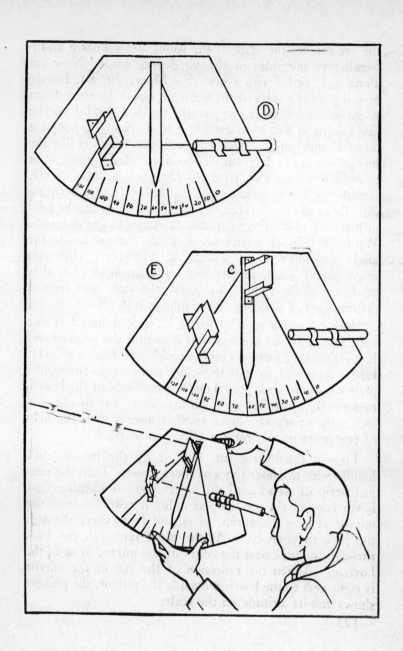

the arc crosses the edge of the wood, draw lines *a* and *b* parallel to the sides of the 60 degree angle. Now cut along *a, b,* and *d* with a saw (See *Figure 59, B*). Locate *m* and *n* on the sides of the angle, at equal distances from *c.* At *m* draw a short line parallel to line *b,* and draw the line joining *m* and *n,* continuing it to *k.* Now cut out of a piece of sheet metal, from a tin can, two brackets like the one shown in the diagram. These are nailed in place at *m* to hold a mirror parallel to *b.* On the extension of *mk,* outside the angle, mount a piece of half-inch pipe pointing directly at the mirror (See *Figure 59, C*). It can be held in place by two strips of metal, as shown in the diagram. With the help of a protractor divide the arc inside the angle into intervals of 5 degrees. However, when you draw in the scale, double the actual measures, so that while the angle is only 60 degrees, the scale runs from 0 degrees to 120 degrees, from right to left.

Now cut out a strip of wood 1″ wide, pointed at one end, and nail it to the board at *c* so that the pointed end just reaches the numbers on the scale (See *Figure 59, D*). Use a long nail so that the same nail can go through a piece of broom stick placed on the other side of the board, to serve as a handle. At *c,* on the center line of the one-inch strip of wood, mount another mirror with the help of two more metal brackets (See *Figure 59, E*).

To use the instrument, hold it by the broom stick handle with the board in a vertical plane. Turn the one-inch strip of wood until the object you are sighting (the lower edge of the sun, for example) is reflected from the mirror at *c* to the mirror at *m* and from there through the pipe to your eye. At the same time you can look through the pipe past the edges of the mirror at *m* to the horizon. When the reflection of the sun in the mirror is right next to the horizon outside the mirror, the pointer shows you its altitude on the scale.

Most of us will never be navigators of ships or planes, and so we will rarely have any need to use our knowledge of navigation. But we will often find it useful on land to be able to find our bearings by locating north. There are several very easy ways to do it.

On a clear night you would locate north by looking for *Polaris,* the *Pole Star.* The pointers of the *Big Dipper* point to the *Pole Star.* It is about half way between the *Big Dipper* and *Cassiopeia* (See pgs. 26 and 27, *Figure 8* and *Figure 11*).

In the daytime you can't see *Polaris,* but you can count on the sun to help you find the north.

Every hiker should know how to make use of a watch for this purpose. Hold your watch level, and place a thin twig, held straight up, over the center of the face of the watch. Now turn the watch until the hour hand coincides with the shadow of the twig. North will be half way between the shadow and the 12 (See *Figure 50*).

If you have no watch, you can use the shadow of a stick in another way. Drive a stick vertically into the

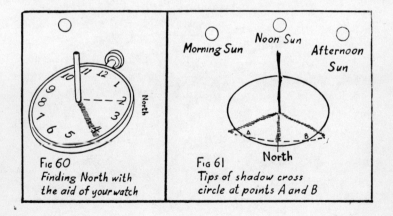

Fig 60
Finding North with
the aid of your watch

Fig 61
Tips of shadow cross
circle at points A and B

ground. As the sun crosses the sky in the course of the day, the shadow of the stick will turn. It will also grow shorter in the morning, and longer again in the afternoon. When the shadow is shortest (close to noon) it will point north.

Another more accurate method requires watching the length of the shadow for a longer time. Draw a circle on the ground, using a stick as center. You can make the circle by tying one end of a rope around the stick, and knotting the other end to a small twig. Then, while pulling the rope tight, swing the extended rope around the center stick and scratch a full circle on the ground with the twig, in the same way that you use a compass. If the circle is not made too small, the end of the shadow of the center stick will be outside the circle in the morning and afternoon, and inside the circle during the middle of the day. There will be two instants when the end of the shadow will be right on the circle. Watch for these instants and draw lines where the shadow rests on the ground. North will be halfway between these two lines (See *Figure 51*).

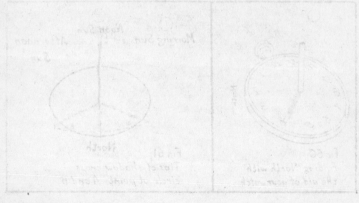

X

THE UNIVERSE OF STARS

THE MILKY WAY IS AN ISLAND UNIVERSE

IF YOU look at the sky in the country, away from the glare of city lights, you can see many more stars than are seen in the city. You will notice how a great number of them bunch together in a band that crosses the sky. This band of starlight is called the Milky Way. It includes the constellations, *Cassiopeia, Perseus, Cygnus, Aquila, Scorpius, Saggitarius,* and the *Southern Cross.*

The Milky Way is the galaxy, or family of stars, to which our sun belongs. It is made up of about a hundred billion stars. It is shaped like a thin flattened disc 100,000 light years in diameter and 10,000 light years wide at its center. Our solar system is located about half-way between the center and the furthest edge of the disc. The sun is moving in this family of stars, and we move with it, towards the bright star *Vega,* in the constellation *Lyra.* The galaxy is like an island in the ocean of space that surrounds it, and is often called an "island universe."

There are many galaxies beside our own Milky Way, extending into outer space.

OUR NEAREST STAR NEIGHBORS

The star closest to us is in the constellation *Centaurus,* and is called *Proxima.* It is 4.3 light-years, or twenty-five thousand billion miles away. *Sirius,* the brightest star we can see in the sky, is the fifth closest star to us, being only 8.8 light-years away. Another close neighbor is the white dwarf known as *Van Maanen's Star* (See pg. 129). Its distance is 12.8 light-years.

CEPHEIDS—THE YARDSTICK OF THE SKY

In the northern sky, near *Cassiopeia,* there is a faint constellation called *Cepheus.* One of its stars, *Delta,* changes in brightness all the time, growing brighter and dimmer, with a regular rhythm. It repeats the same cycle

125

of changes every four days. The reason for these changes in brightness is that the star pulsates, first growing larger, and then contracting again. The complete cycle takes four days. Many pulsating stars like it have been found in our own galaxy, and in other galaxies as well. Because *Delta Cephei* was the first to be studied, these throbbing stars are now known as *Cepheids*.

The *Cepheids* are very important because they serve as the yardstick that helps us measure how far away other galaxies are. When a star throbs, the time it takes for one complete cycle of pulsation is a clue to how bright it really is. This information, combined with how bright the star looks, helps us to find out how far away it is. The further away a star is, the fainter it appears. To measure the distance of a galaxy, the astronomers look for the *Cepheids* in the galaxy and then times the rhythm of their throbbing. This is how they found out that the *Great Nebula* in *Andromeda* is two-million light years away.

THE 200 INCH TELESCOPE

Our astronomers are already reaching into outer space with giant telescopes that are capable of gathering more light from each star. The most powerful of these is the two hundred inch telescope on Mt. Palomar. It has been in use only since 1949, and has already revealed many new and important facts. For example, the scale of distances originally obtained with the help of the *Cepheid* yardstick was too small. It was found that all distances of the outer galaxies are double what they were thought to be before.

XI

COLORS AND BLACKOUTS IN THE SKY

THE COLORS IN STARLIGHT

THE LIGHT we get from the sun is a mixture of colors. You can separate daylight into colors by passing the light through a glass prism. Sometimes droplets of rain do this and spread the colors out in the form of a rainbow. You may also see the colors in your bathroom when sunlight is reflected from the beveled edge of your bathroom mirror. Starlight, too, is a mixture of colors that can be separated by passing the light through a prism.

Another way of separating light into colors is to reflect it from a grating. A grating is a smooth flat surface on which thousands of fine parallel lines have been scratched very close to each other. You can see how this works by looking at light reflected from an LP phonograph record. The micro-groove lines serve as a grating and break up the sunlight or lamplight into colors. When light is separated into its colors, the rainbow-like result is called a *spectrum*.

FINGERPRINTS OF THE ELEMENTS

When a mixture of light passes through a gas that is cooler than the light source, the resulting spectrum is found to be crossed by dark lines. The dark lines represent missing colors that were taken out of the light by the gas. Each chemical element in the gas selects special colors to withdraw, and so, can be recognized by the dark lines it leaves in the spectrum. It's as though each element puts its fingerprint on the light as it passes through. When a detective finds a fingerprint he can identify the person who made it. When an astronomer sees the dark lines in the spectrum of a star he can identify the chemical elements on the surface of the star. This is

127

how we found out that the sun and the most distant stars are all made of the same chemical elements.

THE DISCOVERY OF HELIUM

Scientists compared the dark lines in the spectrum of the sun with dark lines made by known chemicals in the laboratory. They were able to identify the chemicals in the sun, and account for most of the lines. However, there was one set of lines in the sun's spectrum that didn't belong to any known element on earth. This led to the discovery of a new kind of gas on the sun, that was named *Helium* (from the Greek word for sun, *Helios*). It was only after Helium was discovered on the sun that it was found on the earth, too. Helium is a lighter-than-air gas that cannot burn, and is used to inflate dirigibles.

TAKING A STAR'S TEMPERATURE

If you want to take the temperature of the air, you hang a thermometer in it. If you want to take the temperature of a star, billions of miles away, you examine its color. When a piece of iron is heated, it begins to glow. First it will be red-hot. Then, if you make it hot enough, you can get it to be white-hot. Stars, like the iron, glow with different colors, depending on their temperature. In this way, astronomers can measure the surface temperature of the stars. For accurate investigations they will, of course, use instruments to examine the colors in the spectrum of a star. You can, however, see the color differences of stars even with the naked eye. *Vega,* the brightest star in *Lyra,* is bluish-white. *Arcturus,* in Boötes, is golden yellow. *Spica,* in *Virgo,* is pure white, while *Antares,* in *Sorpius,* is distinctly red.

RED GIANTS AND WHITE DWARFS

Stars giving out lots of light are called giants. Stars giving out little light are called dwarfs. These names refer to brightness, not size. But when brightness is combined with color, it does give a clue to size. If a star

128

is very bright and also red, it is a very big star or *red giant*. If it is faint and white, it is a very small star or *white dwarf*.

Antares is a red giant. Its diameter is about 400 million miles. This is so big that if its center were where our sun is, we would be inside the star, half-way between the center and the surface. Its density is so low that it is only 1/3000 that of the air we breathe. On the earth we would call a gas with such a low density a vacuum!

Van Maanen's Star is a white dwarf. Its diameter is only three-fourths the diameter of the earth. But its mass is packed so tight in that small space that its density is 400,000 times as great as the density of water. One cubic inch of it weights 7¼ tons.

THE RED SHIFT

When the spectrum of a distant galaxy is examined it shows familiar lines, by means of which the astronomer identifies the chemicals in it. But they also note that the lines are not exactly in that position in the spectrum, where they would expect to find them. It is as though the whole spectrum had been moved a bit toward the red end of the spectrum. This "red shift" is found to be true of all the galaxies that are far away from us. From this red shift the astronomers conclude that the galaxies are moving away from us, and that the further away a galaxy is, the faster it is moving away. The greatest speed found in this way is 38,000 miles a second, or about one-fifth the speed of light.

You can see that the colors in starlight are a gold-mine of information. They help us learn about the chemical composition, the temperature, the size, and even the speed of the stars.

SPACE CLOUDS

In some parts of the Milky Way, there are dark patches where you might expect to find many stars. These dark patches are clouds of dust in space that block off many stars from view. For a long time they made it impossible to learn anything about the stars that lay behind them.

PENETRATING THE CLOUDS

However, a new method has recently been developed that helps us to "see" through these dust clouds. The secret of the new method is that while the dust clouds stop the *light* of stars from passing through, they do not stop *radio waves*. And, the stars are like radio broadcasting stations, sending out radio waves that can be tuned in by special radio receivers. In this new *radio astronomy* a new kind of telescope is used that is like a radar antenna. (See photograph on pg. 109).

MAPPING THE CLOUDS

Radio astronomy has also made it possible to study the dust clouds themselves, because the hydrogen molecules in the clouds are also sending out radio waves, at a frequency of 1420 megacycles. The study of these radio signals has already led to important information about the distances, position, shapes, and movement of the dust clouds. One conclusion obtained from this information is that the Milky Way *does* have a spiral structure like the spiral nebulae of outer space.

EVOLUTION OF STARS

Now that we can study space dust as well as the stars, we are one step closer to learning how stars came into existence. Most astronomers believe that galaxies were formed out of dust clouds whose particles were brought together by gravitation. As they packed closer together, colliding with each other more and more, the temperature began to go up, until they were hot enough to glow. Astronomers think that the many shapes of galaxies represent different stages in their development. The study of dust clouds will undoubtedly help us to understand this process better.